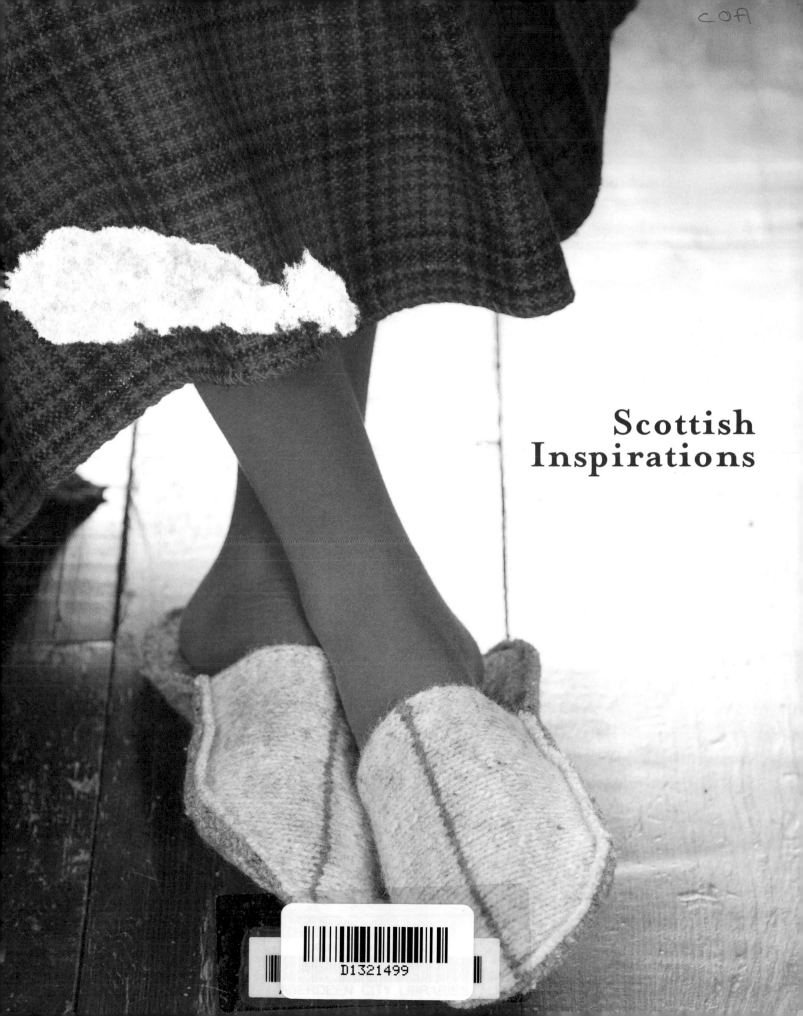

Scottish
Inspirations

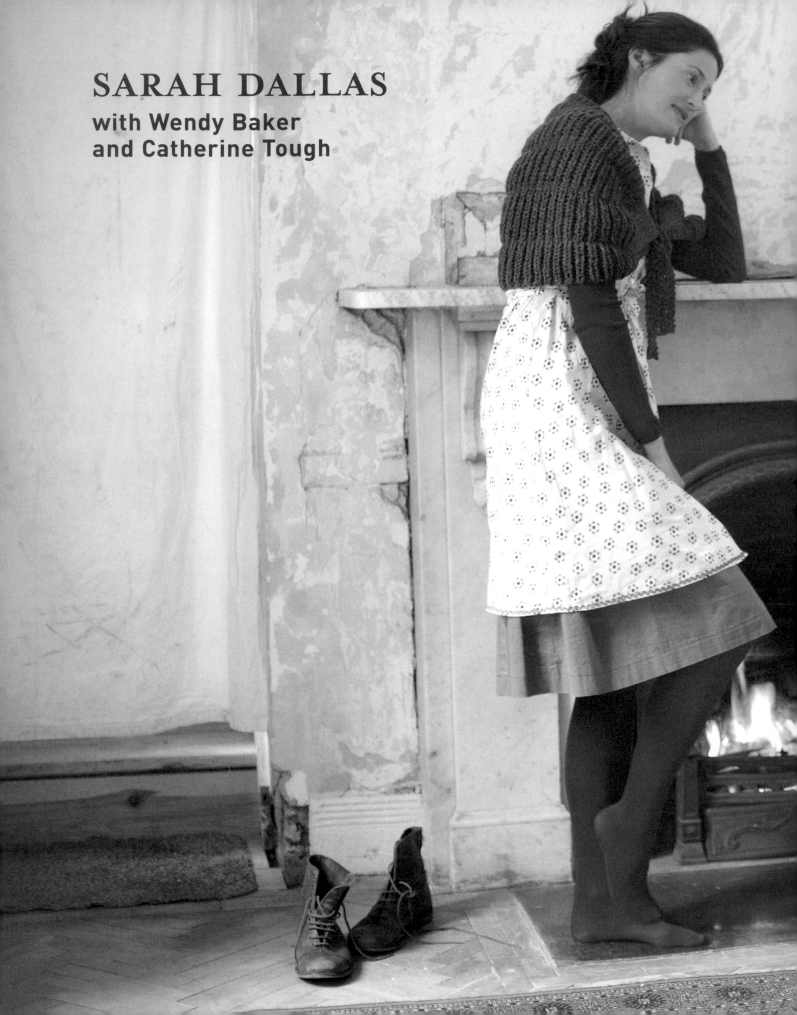

SARAH DALLAS

with Wendy Baker
and Catherine Tough

Scottish inspirations

ROWAN

SCOTTISH INSPIRATIONS

First published by Rowan Yarns in 2006
Rowan Yarns
Green Lane Mill
Holmfirth
West Yorkshire
HD9 2DX

Editor Susan Berry
Designer Anne Wilson
Photographer John Heseltine
Stylist Emma Freemantle
Pattern writers Sue Whiting, Penny Hill and Eva Yates
Pattern checkers Stella Smith and Marilyn Wilson
Diagrams Stella Smith

British Library Cataloguing in Publication Data
A catalogue record of this book is available from the
British Library

ISBN 1-904485-71-5

Reproduced and printed in Singapore

Contents

Introduction

I have enjoyed the opportunity to collaborate with two colleagues on this book, Wendy Baker and Catherine Tough, both of them former students of mine in the School of Fashion Textiles at the Royal College of Art, where I am senior tutor, and both now established designers in their own right. Wendy and Catherine work in a similar way to me. They love subtlety of colour, clarity and simplicity of design, and attention to detail.

We also share a desire to make the most of the natural qualities of the yarn, in this case Rowan's *Scottish Tweed*. Inspired by the hand-dyed yarns of the Scottish Islands, it is robust yet subtle, with flecks of colour and soft shades. The influence of nature can be seen in the colour palette we have chosen: the greys and blacks of stones, the greens and browns of trees, wood and leaves, with the odd flash of colour, such as a heathery mauve.

The Fair Isle designs of Scotland need no introduction to knitters, but we chose in this book to give tradition a contemporary twist. The Fair Isles here are monochromatic – greys, whites and blacks – and very simple. Just a hint of pattern here and there, to bring to life more simple pieces. We also created designs that exploit the great texture of this pure wool yarn, choosing stitches such as cables and moss stitch.

Modern life makes particular demands on us all, and more and more we are looking for a balance between the frenetic urban environment and a desire for the slower pace of the countryside. So the pieces that we have created for this book reflect contemporary lifestyle needs: they are modern in style and shape, yet comforting and relaxing to touch and wear.

We have included a variety of knits: for men, women and children, and for the home. I hope you enjoy knitting them as much as we enjoyed creating them!

SARAH DALLAS

About the designs

The following is a visual reference to the designs created by the three designers, Sarah Dallas, Wendy Baker and Catherine Tough.

SARAH DALLAS

Moss stitch jacket
Rowan *Scottish Tweed Chunky*
Pages 10-14

Lace vest
Rowan *Scottish Tweed 4 ply*
Pages 20-23

Wrap cardigan
Rowan *Scottish Tweed 4 ply*
Pages 24-29

Fair Isle scarf
Rowan *Scottish Tweed 4 ply*
Pages 30-33

Fair Isle stripe throw
Rowan *Scottish Tweed DK*
Pages 34-37

Fair Isle gloves
Rowan *Scottish Tweed 4 ply*
Pages 38-41

Fair Isle stripe cushion
Rowan *Scottish Tweed DK*
Pages 52-55

Chunky stripe cushion
Rowan *Scottish Tweed Chunky*
Pages 56-59

Fair Isle socks
Rowan *Scottish Tweed 4 ply*
Pages 70-73

Cabled blanket coat
Rowan *Scottish Tweed DK*
Pages 74-78

Man's Fair Isle sweater
Rowan *Scottish Tweed DK*
Pages 16-19

Man's cabled sweater
Rowan *Scottish Tweed Chunky*
Pages 46-51

Tie shrug
Rowan *Scottish Tweed 4 ply*
Pages 82-84

Textured scarf
Rowan *Scottish Tweed 4 ply*
Pages 80-81

Rever jacket
Rowan *Scottish Tweed Aran*
Pages 86-90

CATHERINE TOUGH

Felted slippers
Rowan *Scottish Tweed DK*
Pages 42-45

Patchwork throw
Rowan *Scottish Tweed DK*
Pages 60-65

Child's jacket
Rowan *Scottish Tweed DK and 4 ply*
Pages 66-69

Moss stitch jacket

SARAH DALLAS

Sizes

	XS-S	M-L	XL-XXL	
To fit bust	81-86	91-97	102-107	cm
	32-34	36-38	40-42	in
Finished measurements				
Around bust	107	118	129	cm
	42	46½	50¾	in
Length to back neck	54	56	58	cm
	21¼	22	22¾	in
Sleeve seam	37	38	39	cm
	14½	15	15½	in

Yarns

9 (10: 11) x 100g/3½oz balls of Rowan *Scottish Tweed Chunky* in main colour **MC** (Lewis Grey 007) and one ball in **A** (Midnight 023)

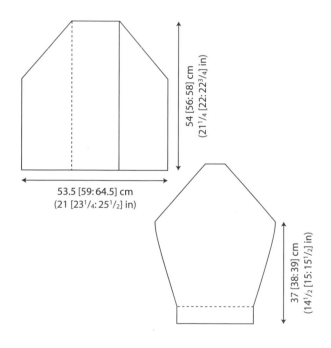

53.5 [59:64.5] cm
(21 [23¼:25½] in)

54 [56:58] cm
(21¼ [22:22¾] in)

37 [38:39] cm
(14½ [15:15½] in)

Needles

Pair of 8mm (UK no 0) (US size 11) knitting needles

Tension

11 sts and 19 rows to 10cm/4in measured over moss st using 8mm (US size 11) needles *or size to obtain correct tension.*

Abbreviations

See page 93.

BACK

Using 8mm (US size 11) needles and A, cast on 59 (65: 71) sts.

Break off A and join in MC.

Row 1 (RS) K1, *P1, K1; rep from * to end.

Row 2 As row 1.

These 2 rows form moss st.

Work in moss st until Back measures 33 (34: 35)cm/13 (13¼: 13¾)in from cast-on edge, ending with RS facing for next row.

Shape raglan armholes

Place markers at both ends of last row to denote base of armholes.

M-L and XL-XXL sizes only

Next row (RS) K1, skpo, moss st to last 3 sts, K2tog, K1.

Next row P1, P2tog, moss st to last 3 sts, P2tog tbl, P1. (61: 67) sts.

Rep last 2 rows (0: 2) times more. (61: 59) sts.

All sizes

Next row (RS) K1, skpo, moss st to last 3 sts, K2tog, K1. 57 (59: 57) sts.

Next row P2, moss st to last 2 sts, P2.

Rep last 2 rows 19 (19: 18) times more. 19 (21: 21) sts.

Break off yarn and leave sts on a holder.

LEFT FRONT

Using 8mm (US size 11) needles and A, cast on
39 (42: 45) sts.

Break off A and join in MC.

Row 1 (RS) *K1, P1; rep from * to last 1 (0: 1) st,
K1 (0: 1).

Row 2 K1 (0: 1), *P1, K1; rep from * to end.

These 2 rows form moss st.

Work in moss st until Left Front matches Back to start
of raglan armhole shaping, ending with RS facing for
next row.

Shape raglan armhole

Place markers at end of last row to denote base of
armholes.

Working all raglan armhole decreases in same way as
for Back, dec 1 st at marked edge of next 1 (3: 7) rows,
then on foll 19 (19: 18) alt rows. 19 (20: 20) sts.

Work 1 row, ending with RS facing for next row.

Break off yarn and leave sts on a holder.

RIGHT FRONT

Using 8mm (US size 11) needles and A, cast on
39 (42: 45) sts.

Break off A and join in MC.

Row 1 (RS) K1 (0: 1), *P1, K1; rep from * to end.

Row 2 *K1, P1; rep from * to last 1 (0: 1) st, K1 (0: 1).

These 2 rows form moss st.

Complete to match Left Front, reversing shapings.

Do NOT break off yarn – set this ball of yarn to one
side as it will be used for Collar.

SLEEVES

Using 8mm (US size 11) needles and A, cast on
31 (33: 35) sts.

Break off A and join in MC.

Work in moss st as given for Back, inc 1 st at each end
of 19th and every foll 6th row to 43 sts, then on every
foll 8th row until there are 49 (51: 53) sts, taking inc
sts into moss st.

Work straight until Sleeve measures 43 (44: 45)cm/
17 (17¼: 17¾)in from cast-on edge, ending with RS
facing for next row.

Shape raglan

Place markers at both ends of last row to denote base
of armholes.

Working all raglan decreases in same way as for Back
raglan armhole, dec 1 st at each end of next and every
foll alt row until 9 sts rem.

Work 1 row, ending with RS facing for next row.

Break off yarn and leave sts on a holder.

MAKING UP

Press lightly on WS following instructions on yarn
label.

Sew raglan seams.

Collar

With RS facing, using 8mm (US size 11) needles and
ball of MC left with Right Front, work across Right
Front sts as folls: moss st 16 (17: 17) sts, K2tog, K1,
place marker on needle, work across sts of Right
Sleeve as folls: K1, skpo, moss st 3 sts, K2tog, K1,
place marker on needle, work across sts of Back as
folls: K1, skpo, moss st 13 (15: 15) sts, K2tog, K1,
place marker on needle, work across sts of Left Sleeve
as folls: K1, skpo, moss st 3 sts, K2tog, K1, place
marker on needle, then work across Left Front sts as
folls: K1, skpo, moss st to end. 67 (71: 71) sts.

Row 1 (WS) [Moss st to within 2 sts of marker, P1,
K2tog and move marker onto this st, P1] 4 times, moss
st to end. 63 (67: 67) sts.

Row 2 [Moss st to within 1 st of marked st, sl 1, K2tog
(marked st is first of these 2 sts), psso] 4 times, moss
st to end. 55 (59: 59) sts.

Work in moss st as set for 13 rows, ending with RS
facing for next row.

Row 16 Moss st 15 (16: 16) sts, sl 1, K2tog, psso, moss
st 19 (21: 21) sts, sl 1, K2tog, psso, moss st to end. 51
(55: 55) sts.

Work in moss st for 3 rows more, ending with RS
facing for next row.

Cast off in moss st.

Sew side and sleeve seams, reversing sleeve seam for
first 7cm/2¾in for turn-back. Fold 6cm/2¼in cuff
to RS.

Man's Fair Isle sweater

WENDY BAKER

Sizes

	S-M	L-XL	XXL-XXXL	
To fit chest	97-102	107-112	117-122	cm
	38-40	42-44	46-48	in
Finished measurements				
Around chest	107	118	129	cm
	42¼	46½	50¾	in
Length to shoulder	64	66	68	cm
	25¼	26	26¾	in
Sleeve seam	52	53	54	cm
	20½	20¾	21¼	in

Yarns

Rowan *Scottish Tweed DK*:

MC Storm Grey 004   12 (14: 15) x 50g/1¾oz balls
A   Lewis Grey 007   1 (1: 1) x 50g/1¾oz balls
B   Grey Mist 001   1 (1: 1) x 50g/1¾oz balls

Needles

Pair of 3¾mm (UK no 9) (US size 5) knitting needles
Pair of 4mm (UK no 8) (US size 6) knitting needles

Tension

22 sts and 30 rows to 10cm/4in measured over st st using 4mm (US size 6) needles *or size to obtain correct tension.*

Abbreviations

See page 93.

Special abbreviation

Tw2 = K2tog leaving sts on left needle, K first st again and slip both sts off left needle together

Special note

When working patt in st st from chart, strand yarn not in use loosely across WS of work, weaving it in every 3 or 4 sts. Work odd-numbered rows as K rows, reading them from right to left, and even-numbered rows as P rows, reading them from left to right.

BACK

Using 3¾mm (US size 5) needles and A, cast on 118 (130: 142) sts.
Row 1 (RS) P2, *K2, P2; rep from * to end.
Row 2 K2, *P2, K2; rep from * to end.
Row 3 P2, *Tw2, P2; rep from * to end.
Row 4 As row 2.
These 4 rows form fancy rib.
Break off A and join in MC.
Work in fancy rib for 24 rows more, dec 1 st at end of last row and ending with RS facing for next row.
117 (129: 141) sts.
Change to 4mm (US size 6) needles.
Starting with a K row, work in st st until Back measures 41 (42: 43)cm/16 (16½: 17)in from cast-on edge, ending with RS facing for next row.

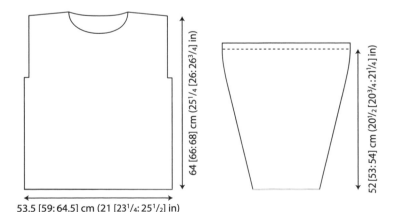

64 [66:68] cm (25¼ [26:26¾] in)

53.5 [59:64.5] cm (21 [23¼: 25½] in)

52[53:54] cm (20½ [20¾:.21¼] in)

Shape armholes

Cast off 6 sts at beg of next 2 rows. 105 (117: 129) sts.

Work straight until armhole measures 10 (11: 12)cm/
4 (4¼: 4¾)in, ending with RS facing for next row.

Place chart

Starting and ending rows as indicated, now work patt
in st st from chart, working chart rows 1 to 42 once
and then completing work in st st using MC only if
required, as folls:

Work straight until armhole measures 21 (22: 23)cm/
8¼ (8¾: 9)in, ending with RS facing for next row.

Shape back neck

Next row (RS) Patt 34 (39: 43) sts and turn, leaving
rem sts on a holder.

Work each side of neck separately.

Keeping patt correct, dec 1 st at neck edge of next 4
rows. 30 (35: 39) sts.

Work 1 row, ending with RS facing for next row.

Shape shoulder

Cast off 10 (12: 13) sts at beg of next and foll alt row.

Work 1 row.

Cast off rem 10 (11: 13) sts.

With RS facing, rejoin yarn to rem sts, cast off centre
37 (39: 43) sts, patt to end.

Complete to match first side, reversing shapings.

FRONT

Work as given for Back until 18 (20: 22) rows less have
been worked than on Back to start of shoulder
shaping, ending with RS facing for next row.

Shape neck

Next row (RS) Patt 44 (50: 55) sts and turn, leaving
rem sts on a holder.

Work each side of neck separately.

Keeping patt correct, dec 1 st at neck edge of next 13
rows, then on foll 1 (2: 3) alt rows. 30 (35: 39) sts.

Work 2 rows, ending with RS facing for next row.

Shape shoulder

Cast off 10 (12: 13) sts at beg of next and foll alt row.

Work 1 row.

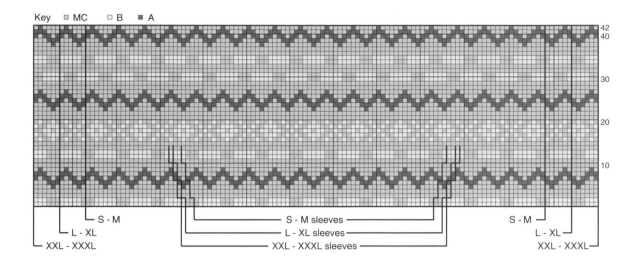

Key ☐ MC   ☐ B   ■ A

Cast off rem 10 (11: 13) sts.

With RS facing, rejoin yarn to rem sts, cast off centre 17 (17: 19) sts, patt to end.

Complete to match first side, reversing shapings.

SLEEVES

Using 3¾mm (US size 5) needles and A, cast on 54 (58: 62) sts.

Work in fancy rib as given for Back for 4 rows.

Break off A and join in MC.

Work in fancy rib for 16 rows more, inc (inc: dec) 1 st at end of last row and ending with RS facing for next row. 55 (59: 61) sts.

Change to 4mm (US size 6) needles.

Place chart

Starting and ending rows as indicated, now work in patt from chart as folls:

Inc 1 st at each end of 3rd and 2 foll 4th rows. 61 (65: 67) sts.

Work 3 rows, ending after chart row 14 and with RS facing for next row.

Break off contrasts and cont using MC **only**.

Starting with a K row, work in st st, shaping sides by inc 1 st at each end of next and every foll 4th row to 89 (91: 95) sts, then on every foll 6th row until there are 107 (111: 115) sts.

Work straight until Sleeve measures 52 (53: 54)cm/ 20½ (20¾: 21¼)in from cast-on edge, ending with RS facing for next row.

Shape top

Place markers at both ends of last row.

Work 8 rows more, ending with RS facing for next row.

Cast off.

MAKING UP

Press lightly on WS following instructions on yarn label and avoiding ribbing.

Sew right shoulder seam.

Collar

With RS facing, using 3¾mm (US size 5) needles and MC, pick up and knit 18 (20: 22) sts down left side of neck, 17 (17: 19) sts from Front, 18 (20: 22) sts up right side of neck, then 57 (57: 59) sts from Back. 110 (114: 122) sts.

Starting with a P row, work in st st until Collar measures 6cm/2¼in from pick-up row, ending with **WS** of body facing for next row.

Starting with row 1 (to reverse RS of work), now work in fancy rib as given for Back for 16 rows, ending with RS of Collar (WS of body) facing for next row.

Break off MC and join in A.

Work in fancy rib for 4 rows more.

Cast off in patt.

Sew left shoulder and Collar seam, reversing collar seam for turn-back. Matching sleeve markers to top of side seams and centre of sleeve cast-off edge to shoulder seams, sew Sleeves into armholes. Sew side and sleeve seams.

Lace vest

SARAH DALLAS

Sizes

XS	S	M	L	XL	XXL	
To fit bust						
81	86	91	97	102	107	cm
32	34	36	38	40	42	in
Finished measurements						
Around bust						
79	85	92	98	104	110	cm
31	33½	36¼	38½	41	43¼	in
Length to shoulder						
51	52	53	54	55	56	cm
20	20½	20¾	21¼	21½	22	in

Yarns

5 (6: 6: 7: 8: 8) x 25g/⅞oz balls of Rowan *Scottish Tweed 4 ply* in main colour **MC** (Porridge 024) and one ball each in **A** (Lewis Grey 007) and **B** (Grey Mist 001)

Needles

Pair of 3mm (UK no 11) (US size 3) knitting needles
Pair of 3¾mm (UK no 9) (US size 5) knitting needles
3mm (UK no 11) (US size 3) circular knitting needle
3.50mm (UK no 9) (US size E-4) crochet hook

Extras

120cm/48in of narrow ribbon

Tension

26 sts and 32 rows to 10cm/4in measured over patt using 3¾mm (US size 5) needles *or size to obtain correct tension.*

Abbreviations

See page 93.

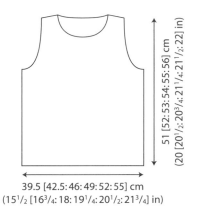

51 [52: 53: 54: 55: 56] cm
(20 [20½: 20¾: 21¼: 21½: 22] in)

39.5 [42.5: 46: 49: 52: 55] cm
(15½ [16¾: 18: 19¼: 20½: 21¾] in)

BACK

Using 3mm (US size 3) needles and A, cast on 103 (111: 119: 127: 135: 143) sts.

Break off A and join in MC.

Rib row 1 (RS) K1, *P1, K1; rep from * to end.

Rib row 2 P1, *K1, P1; rep from * to end.

These 2 rows form rib.

Work in rib for 9 rows more, ending with **WS** facing for next row.

Change to 3¾mm (US size 5) needles.

Now work in patt as folls:

Row 1 (WS) K2, P3, K2, *P1, K2, P3, K2; rep from * to end.

Row 2 P2, yon, sl 1, K2tog, psso, yfrn, P2, *K1, P2, yon, sl 1, K2tog, psso, yfrn, P2; rep from * to end.

Row 3 As row 1.

Row 4 P2, K1, yfwd, skpo, P2, *K1, P2, K1, yfwd, skpo, P2; rep from * to end.

Row 5 As row 1.

Row 6 P2, K3, P2, *K1, P2, K3, P2; rep from * to end.

These 6 rows form patt.

Cont in patt until Back measures 33 (34: 34: 35: 35: 36)cm/13 (13¼: 13¼: 13¾: 13¾: 14)in, ending with RS facing for next row.

Shape armholes

Keeping patt correct, cast off 4 (5: 5: 6: 6: 7) sts at beg of next 2 rows. 95 (101: 109: 115: 123: 129) sts.

Dec 1 st at each end of next 5 (7: 9: 11: 13: 15) rows, then on foll 7 alt rows. 71 (73: 77: 79: 83: 85) sts.**

Work straight until armhole measures 18 (18: 19: 19: 20: 20)cm/7 (7: 7½: 7½: 7¾: 7¾)in, ending with RS facing for next row.

Shape back neck and shoulders

Next row (RS) Cast off 4 (4: 3: 3: 5: 5) sts, patt until there are 14 (14: 17: 17: 17: 18) sts on right needle and turn, leaving rem sts on a holder.

Work each side of neck separately.

Keeping patt correct, cast off 2 sts at beg of next row, 3 (3: 4: 4: 4: 5) sts at beg of foll row, 2 sts at beg of next row, then 3 (3: 4: 4: 4: 4) sts at beg of foll row.

Dec 1 st at neck edge of next row.

Cast off rem 3 (3: 4: 4: 4: 4) sts.

With RS facing, rejoin yarn to rem sts, cast off centre 35 (37: 37: 39: 39: 39) sts, patt to end.

Complete to match first side, reversing shapings.

FRONT

Work as given for Back to **.

Work 3 (1: 3: 1: 3: 1) rows, ending with RS facing for next row.

Shape neck

Next row (RS) Patt 28 (28: 30: 30: 32: 33) sts and turn, leaving rem sts on a holder.

Work each side of neck separately.

Keeping patt correct, dec 1 st at neck edge of next 8 rows, then on foll 7 alt rows. 13 (13: 15: 15: 17: 18) sts.

Work straight until Front matches Back to start of shoulder shaping, ending with RS facing for next row.

Shape shoulder

Cast off 4 (4: 3: 3: 5: 5) sts at beg of next row, 3 (3: 4: 4: 4: 5) sts at beg of foll alt row, then 3 (3: 4: 4: 4: 4) sts at beg of foll alt row.

Work 1 row.

Cast off rem 3 (3: 4: 4: 4: 4) sts.

With RS facing, rejoin yarn to rem sts, cast off centre 15 (17: 17: 19: 19: 19) sts, patt to end.

Complete to match first side, reversing shapings.

MAKING UP

Press lightly on WS following instructions on yarn label and avoiding ribbing.

Sew shoulder seams.

Neckband

With RS facing, using 3mm (US size 3) circular needle and MC, starting and ending at left shoulder seam, pick up and knit 34 (33: 33: 32: 32: 32) sts down left side of neck, 16 (16: 16: 18: 18: 18) sts from Front, 34 (33: 33: 32: 32: 32) sts up right side of neck, then 56 (58: 58: 58: 58: 58) sts from Back. 140 sts.

Rounds 1 and 2 Knit.

Round 3 K3 (2: 2:2: 2: 2), *yfwd, K2tog, K5; rep from * to last 4 (5: 5: 5: 5: 5) sts, yfwd, K2tog, K2 (3: 3: 3: 3: 3).

Round 4 and 5 Knit.

Cast off knitwise.

Neck edging

With RS facing, using 3.50mm (US size E-4) crochet hook and B, join yarn with a ss to cast-off edge of Neckband above one shoulder seam and work around neck edge as folls: 1 ch (does NOT count as st), 1 dc into cast-off st where yarn was joined on, *5 ch, 1ss into last dc, 1 dc into each of next 4 cast-off sts; rep from * to last 3 cast-off sts, 5 ch, 1ss into last dc, 1 dc into each of last 3 cast-off sts, join with a ss to first dc. Fasten off.

Armhole borders (both alike)

With RS facing, using 3mm (US size 3) needles and MC, pick up and knit 105 (105: 109: 119: 113: 113) sts evenly all round armhole edge.

Work in rib as given for Back for 5 rows, ending with RS facing for next row.

Break off MC and join in B.

Cast off in rib.

Sew side and Armhole Border seams. Starting and ending either side of centre front, thread ribbon through round 3 of Neckband and tie in bow at front.

Wrap cardigan

SARAH DALLAS

Sizes

	XS-S	M-L	XL-XXL	
To fit bust	81-86	91-97	102-107	cm
	32-34	36-38	40-42	in
Finished measurements				
Around bust	93	103	113	cm
	36½	40½	44½	in
Length to shoulder	39	42	44	cm
	15½	16½	17¼	in
Sleeve seam	46	47	48	cm
	18	18½	19	in

Yarns

10 (12: 13) x 25g/⅞oz balls of Rowan *Scottish Tweed 4 ply* in main colour **MC** (Lavender 005) and one ball each in **A** (Apple 015) and **B** (Brilliant Pink 010)

Needles

Pair of 3¼mm (UK no 10) (US size 3) knitting needles
Pair of 3¾mm (UK no 9) (US size 5) knitting needles
3.00mm (UK no 10) (US size D3) crochet hook

Tension

24 sts and 32 rows to 10cm/4in measured over patt using 3¾mm (US size 5) needles *or size to obtain correct tension*.

Abbreviations

See page 93.

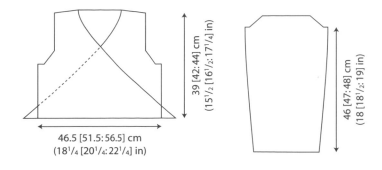

39 [42:44] cm
(15½ [16½:17¼] in)

46.5 [51.5:56.5] cm
(18¼ [20¼:22¼] in)

46 [47:48] cm
(18 [18½:19] in)

BACK

Using 3¼mm (US size 3) needles and A, cast on 111 (123: 135) sts.

Break off A and join in MC.

Rib row 1 (RS) K1, *P1, K1; rep from * to end.

Rib row 2 P1, *K1, P1; rep from * to end.

These 2 rows form rib.

Work in rib for 4 rows more, ending with RS facing for next row.

Change to 3¾mm (US size 5) needles.

Now work in patt as folls:

Row 1 (RS) Knit.

Row 2 and every foll alt row Purl.

Row 3 K3 (1: 7), *yfwd, skpo, K6; rep from * to last 4 (2: 8) sts, yfwd, skpo, K2 (0: 6).

Row 5 K1 (2: 5), [yfwd, skpo, K3] 0 (1: 0) times, *K2tog, yfwd, K1, yfwd, skpo, K3; rep from * to last 6 (4: 2) sts, [K2tog, yfwd] 1 (1: 0) times, [K1, yfwd, skpo] 1 (0: 0) times, K1 (2: 2).

Row 7 As row 3.

Row 9 Knit.

Row 11 K7 (5: 3), *yfwd, skpo, K6; rep from * to last 8 (6: 4) sts, yfwd, skpo, K6 (4: 2).

Row 13 K5 (3: 1), *K2tog, yfwd, K1, yfwd, skpo, K3; rep from * to last 2 (0: 6) sts, [K2tog, yfwd, K1, yfwd, skpo] 0 (0: 1) times, K2 (0: 1).

Row 15 As row 11.

Row 16 As row 2.

These 16 rows form patt.

Cont in patt for 44 (48: 52) rows more, ending with RS facing for next row.

Shape armholes

Keeping patt correct, cast off 5 (6: 7) sts at beg of next 2 rows. 101 (111: 121) sts.

M-L and XL-XXL sizes only

Next row (RS) K2, skpo, patt to last 4 sts, K2tog, K2.

Next row P2, P2tog, P to last 4 sts, P2tog tbl, P2. (107: 117) sts.

XL-XXL size only

Rep last 2 rows once more. 113 sts.

All sizes

Next row (RS) K2, skpo, patt to last 4 sts, K2tog, K2.

Next row Purl.

Rep last 2 rows 8 (9: 10) times more. 83 (87: 91) sts.

Work straight until armhole measures 19 (20: 21)cm/ 7½ (7¾: 8¼)in, ending with RS facing for next row.

Shape shoulders

Keeping patt correct, cast off 5 (5: 6) sts at beg of next 6 (4: 8) rows, then 6 (6: 0) sts at beg of foll 2 (4: 0) rows.

Cast off rem 41 (43: 43) sts.

LEFT FRONT

Using 3¼mm (US size 3) needles and A, cast on 120 (132: 144) sts.

Break off A and join in MC.

Rib row 1 (RS) *K1, P1; rep from * to end.

Rib row 2 As row 1.

These 2 rows form rib.

Work in rib for 4 rows more, ending with RS facing for next row.

Change to 3¾mm (US size 5) needles.

Now work in patt, shaping front slope, as folls:

Row 1 (RS) K to last 3 sts, K2tog, K1. 119 (131: 143) sts.

Row 2 and every foll alt row P1, P2tog, P to end.

Row 3 K3 (1: 7), *yfwd, skpo, K6; rep from * to last 3 (9: 7) sts, [yfwd, skpo] 0 (1: 1) times, K0 (4: 2), K2tog, K1. 117 (129: 141) sts.

Row 5 K1 (2: 5), [yfwd, skpo, K3] 0 (1: 0) times, *K2tog, yfwd, K1, yfwd, skpo, K3; rep from * to last 3 (9: 7) sts, [K2tog, yfwd] 0 (1: 1) times, [K1, yfwd, skpo] 0 (1: 0) times, K0 (1: 2), K2tog, K1. 115 (127: 139) sts.

Row 7 K3 (1: 7), *yfwd, skpo, K6; rep from * to last
7 (5: 11) sts, yfwd, skpo, K2 (0: 6), K2tog, K1.
113 (125: 137) sts.

Row 9 As row 1. 111 (123: 135) sts.

Row 11 K7 (5: 3), *yfwd, skpo, K6; rep from * to last
7 (5: 3) sts, [yfwd, skpo] 1 (1: 0) times, K2 (0: 0), K2tog,
K1. 109 (121: 133) sts.

Row 13 K5 (3: 1), *K2tog, yfwd, K1, yfwd, skpo, K3; rep
from * to last 7 (5: 3) sts, [K2tog, yfwd] 1 (0: 0) times,
K2 (2: 0), K2tog, K1. 107 (119: 131) sts.

Row 15 K7 (5: 3), *yfwd, skpo, K6; rep from * to last
11 (9: 7) sts, yfwd, skpo, K6 (4: 2), K2tog, K1.
105 (117: 129) sts.

Row 16 As row 2. 104 (116: 128) sts.

These 16 rows form patt and start front slope shaping.
Cont in patt, dec 1 st at front slope edge of next
43 (48: 52) rows. 61 (68: 76) sts.

Work 1 (0: 0) row more, ending with RS facing for next
row.

Shape armhole

Keeping patt correct, cast off 5 (6: 7) sts at beg and
dec 1 st at end of next row. 55 (61: 68) sts.

Work 1 row.

Working all armhole decreases as set by Back, dec 1
st at armhole edge of next 1 (3: 5) rows, then on foll
8 (9: 10) alt rows **and at same time** dec 1 st at front
slope edge of next and every foll alt row.
37 (38: 40) sts.

Dec 1 st at front slope edge **only** of every foll alt row
until 21 (22: 24) sts rem.

Work straight until Left Front matches Back to start of
shoulder shaping, ending with RS facing for next row.

Shape shoulder

Keeping patt correct, cast off 5 (5: 6) sts at beg of next
and foll 2 (1: 2) alt rows, then 0 (6: 0) sts at beg of foll
0 (1: 0) alt row.

Work 1 row.

Cast off rem 6 sts.

RIGHT FRONT

Using 3¼mm (US size 3) needles and A, cast on
120 (132: 144) sts.

Break off A and join in MC.

Rib row 1 (RS) *P1, K1; rep from * to end.

Rib row 2 As row 1.

These 2 rows form rib.

Work in rib for 4 rows more, ending with RS facing for
next row.

Change to 3¾mm (US size 5) needles.

Now work in patt, shaping front slope, as folls:

Row 1 (RS) K1, skpo, K to end. 119 (131: 143) sts.

Row 2 and every foll alt row P to last 3 sts, P2tog tbl,
P1.

Row 3 K1, skpo, K7 (5: 3), *yfwd, skpo, K6; rep from *
to last 4 (2: 8) sts, yfwd, skpo, K2 (0: 6).
117 (129: 141) sts.

Row 5 K1, skpo, K3 (1: 2), [yfwd, skpo, K3] 0 (0: 1)
times, *K2tog, yfwd, K1, yfwd, skpo, K3; rep from * to
last 6 (4: 2) sts, [K2tog, yfwd] 1 (1: 0) times, [K1, yfwd,
skpo] 1 (0: 0) times, K1 (2: 2). 115 (127: 139) sts.

Row 7 K1, skpo, K3 (1: 7), *yfwd, skpo, K6; rep from *
to last 4 (2: 8) sts, yfwd, skpo, K2 (0: 6).
113 (125: 137) sts.

Row 9 As row 1. 111 (123: 135) sts.

Row 11 K1, skpo, K3 (1: 7), *yfwd, skpo, K6; rep from *
to last 8 (6: 4) sts, yfwd, skpo, K6 (4: 2).
109 (121: 133) sts.

Row 13 K1, skpo, K2 (0: 0), [yfwd, skpo] 1 (1: 0) times,
K3, *K2tog, yfwd, K1, yfwd, skpo, K3; rep from * to last
2 (0: 6) sts, [K2tog, yfwd, K1, yfwd, skpo] 0 (0: 1) times,
K2 (0: 1). 107 (119: 131) sts.

Row 15 K1, skpo, K7 (5: 3), *yfwd, skpo, K6; rep from *
to last 8 (6: 4) sts, yfwd, skpo, K6 (4: 2).
105 (117: 129) sts.

Row 16 As row 2. 104 (116: 128) sts.

These 16 rows form patt and start front slope shaping.

Complete to match Left Front, reversing shapings.

SLEEVES

Using 3¼mm (US size 3) needles and B, cast on
55 (59: 61) sts.

Break off B and join in MC.

Work in rib as given for Back for 5cm/2in, ending with
RS facing for next row.

Change to 3¾mm (US size 5) needles.

Now work in patt as folls:

Row 1 (RS) Knit.

Row 2 and every foll alt row Purl.

Row 3 K3 (5: 6), *yfwd, skpo, K6; rep from * to last 4 (6: 7) sts, yfwd, skpo, K2 (4: 5).

Row 5 K1 (3: 4), *K2tog, yfwd, K1, yfwd, skpo, K3; rep from * to last 6 (0: 1) sts, [K2tog, yfwd, K1, yfwd, skpo] 1 (0: 0) times, K1 (0: 1).

Row 7 As row 3.

Row 9 Knit.

Row 11 K7 (1: 2), *yfwd, skpo, K6; rep from * to last 8 (2: 3) sts, yfwd, skpo, K6 (0: 1).

Row 13 [Inc in first st] 0 (0: 1) times, K5 (2: 2), [yfwd, skpo, K3] 0 (1: 1) times, *K2tog, yfwd, K1, yfwd, skpo, K3; rep from * to last 2 (4: 5) sts, [K2tog, yfwd] 0 (1: 1) times, K2, [inc in last st] 0 (0: 1) times. 55 (59: 63) sts.

Row 15 [Inc in first st] 1 (1: 0) times, K6 (0: 3), *yfwd, skpo, K6; rep from * to last 8 (10: 4) sts, yfwd, skpo, K5 (7: 2), [inc in last st] 1 (1: 0) times. 57 (61: 63) sts.

Row 16 As row 2.

These 16 rows form patt and start sleeve shaping.

Cont in patt, shaping sides by inc 1 st at each end of 15th (15th: 11th) and every foll 16th (16th: 14th) row to 65 (67: 67) sts, then on every foll 18th (18th: 16th) row until there are 69 (73: 77) sts, taking inc sts into patt. Work straight until Sleeve measures 46 (47: 48)cm/ 18 (18½: 19)in from cast-on edge, ending with RS facing for next row.

Shape top

Keeping patt correct, cast off 5 (6: 7) sts at beg of next 2 rows. 59 (61: 63) sts.

Working all decreases as set by Back armhole decreases, dec 1 st at each end of next and foll 7 (8: 9) alt rows. 43 sts.

Work 1 row, ending with RS facing for next row.

Cast off 2 sts at beg of next 6 rows.

Cast off rem 31 sts.

MAKING UP

Press lightly on WS following instructions on yarn label and avoiding ribbing.

Sew shoulder seams.

Neck edging

With RS facing, using 3.00mm (US size D-3) crochet hook and B, join yarn with a ss to cast-on edge of Right Front at base of front slope shaping.

Working into row-end edges along front slopes and cast-off sts across back neck, work along entire front slope and back neck edge as folls: 1 ch (does NOT count as st), 1 dc into point where yarn was joined on, 1 dc into edge, *3 ch, 1 dc into same place as last dc, 2 dc into edge; rep from * to cast-on edge of Left Front. Fasten off.

Sew side seams. Sew sleeve seams. Sew Sleeves into armholes.

Fair Isle scarf

SARAH DALLAS

Size

The finished scarf measures 32cm/12½in by 138cm/54¼in.

Yarns

5 x 25g/⅞oz balls of Rowan *Scottish Tweed 4 ply* in main colour **MC** (Grey Mist 001) and one ball each in **A** (Sunset 011), **B** (Lewis Grey 007), **C** (Porridge 024), **D** (Midnight 023) and **E** (Brilliant Pink 010)

Needles

Pair of 3¾mm (UK no 9) (US size 5) knitting needles

Tension

24 sts and 32 rows to 10cm/4in measured over st st using 3¾mm (US size 5) needles *or size to obtain correct tension.*

Abbreviations

See page 93.

Special note

When working patt from chart, strand yarn not in use loosely across WS of work, weaving it in every 3 or 4 sts. Work odd-numbered rows as RS rows, reading them from right to left, and even-numbered rows as WS rows, reading them from left to right.

SCARF

Using 3¾mm (US size 5) needles and A, cast on 77 sts. Break off A and join in MC.

Row 1 (RS) [K1, P1] twice, K to last 4 sts, [P1, K1] twice.

Row 2 K1, P1, K1, P to last 3 sts, K1, P1, K1.

These 2 rows form patt.

Work in patt for 6 rows more, ending with RS facing for next row.

Work first Fair Isle band

Joining in and breaking off B, C and D as required, work from chart as folls:

Work chart rows 1 to 12, 3 times, then rep chart rows 1 to 9 again, ending with **WS** facing for next row.

Break off B, C and D and cont using MC only.

Work straight in patt until Scarf measures 132cm/52in from cast-on edge, ending with RS facing for next row.

Work second Fair Isle band

Joining in and breaking off B, C and D as required, work chart rows 1 to 9, ending with **WS** facing for next row.

Break off B, C and D and cont using MC only.

Work straight in patt for 9 rows, ending with RS facing for next row.

Break off MC and join in E.

Next row (RS) Using E, knit.

Cast off purlwise.

MAKING UP

Press lightly on WS following instructions on yarn label.

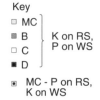

Key
- □ MC
- ■ B } K on RS, P on WS
- □ C
- ■ D
- ▣ MC - P on RS, K on WS

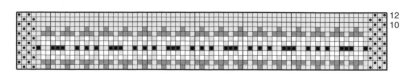

Fair Isle stripe throw

SARAH DALLAS

Size

The finished throw measures 130cm/51in by
172cm/67¾in.

Yarns

Rowan *Scottish Tweed DK*:

A	Grey Mist 001	10 x 50g/1¾oz balls
B	Storm Grey 004	10 x 50g/1¾oz balls
C	Lavender 005	1 x 50g/1¾oz ball
D	Lewis Grey 007	1 x 50g/1¾oz ball
E	Midnight 023	1 x 50g/1¾oz ball
F	Brilliant Pink 010	1 x 50g/1¾oz ball
G	Apple 015	1 x 50g/1¾oz ball

Needles

Pair of 4½mm (UK no 7) (US size 7) knitting needles
4.00mm (UK no 8) (US size G-6) crochet hook

Tension

20 sts and 26 rows to 10cm/4in measured over st st
using 4½mm (US size 7) needles *or size to obtain
correct tension.*

Abbreviations

See page 93.

Special note

When working patt from chart, strand yarn not in use
loosely across WS of work, weaving it in every 3 or 4
sts. Work odd-numbered rows as RS rows, reading
them from right to left, and even-numbered rows as
WS rows, reading them from left to right.

FIRST STRIP

Using 4½mm (US size 7) needles and C, cast on 130 sts.

Break off C and join in A.

Row 1 (RS) Knit.

Row 2 K5, P to end.

These 2 rows form patt.

Work in patt until Strip measures 51cm/20in, ending with RS facing for next row.

Work Fair Isle band

Join in D.

Following chart A, work chart rows 1 to 11, ending with **WS** facing for next row.

Break off D and cont using A only.

Work straight in patt until Strip measures 172cm/67¾in from cast-on edge, ending with **WS** facing for next row.

Break off A and join in F.

Next row (WS) Purl.

Cast off.

SECOND STRIP

Using 4½mm (US size 7) needles and C, cast on 130 sts.

Break off C and join in B.

Row 1 (RS) Knit.

Row 2 P to last 5 sts, K5.

These 2 rows form patt.

Work in patt until Strip measures 118cm/46½in, ending with RS facing for next row.

Work Fair Isle band

Join in E.

Following chart B, work chart rows 1 to 11, ending with WS facing for next row.

Break off E and cont using B only.

Work straight in patt until Strip measures 172cm/67¾in from cast-on edge, ending with **WS** facing for next row.

Break off B and join in F.

Next row (WS) Purl.

Cast off.

MAKING UP

Press lightly on WS following instructions on yarn label.

Join strips

Lay First Strip flat with WS uppermost, then lay Second Strip on top of First, with WS together and cast-on and cast-off edges matching.

Using 4.00mm (US size G-6) crochet hook and G, join Strips by working a row of dc along central row-end edges, working each st through both layers. (Garter st borders of Strips form outer edges.)

Fasten off.

Key

ChartA

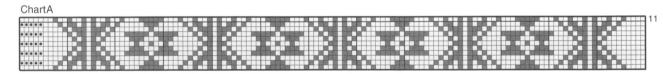

11

Key

ChartB

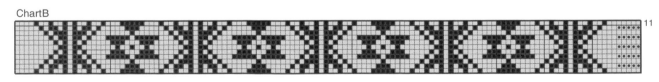

11

Fair Isle gloves

SARAH DALLAS

Size

The finished gloves measure 19cm/7½in around palm of hand.

Yarns

2 x 25g/⅞oz balls of Rowan *Scottish Tweed 4 ply* in main colour **MC** (Grey Mist 001) and one ball each in **A** (Midnight 023), **B** (Lewis Grey 007) and **C** (Porridge 024)

Needles

Set of four 2¾mm (UK no 12) (US size 2) double-pointed knitting needles

Tension

32 sts and 36 rows to 10cm/4in measured over st st using 2¾mm (US size 2) needles *or size to obtain correct tension.*

Abbreviations

See page 93.

Special note

When working patt from chart, strand yarn not in use loosely across WS of work, weaving it in every 3 or 4 sts. Work ALL rounds as RS (knit) rounds, reading them from right to left.

RIGHT GLOVE

Using 2¾mm (US size 2) double-pointed needles and A, cast on 60 sts, distributing them evenly over 3 needles. Break off A and join in MC.

Round 1 (RS) *K1, P1; rep from * to end.

This round forms rib.

Work in rib for 36 rounds more, ending with RS facing for next row.

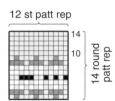

Key
☐ MC
▨ B
☐ C
■ A

12 st patt rep

14 round patt rep

Round 38 [Rib 2, inc in next st, rib 2] 12 times. 72 sts.

Joining in A, B and C as required, repeating the 12-st patt rep 6 times around each round and repeating the 14 round patt rep throughout, start with chart row 1 and work from chart as folls:

Work 24 rounds.**

Shape thumb

Round 25 Slip first 18 sts of next round onto a safety pin, turn and cast on 6 sts, turn and patt rem 54 sts of round. 60 sts.

Keeping patt correct (by now repeating the 12 st patt rep 5 times around each round), work 13 rounds, ending after chart round 10.

Break off A, B and C and cont in rounds of st st (K every round) using MC only.

Work 4 rounds.

Shape first finger

Next round K5, slip next 44 sts onto a holder, turn and cast on 2 sts, turn and K rem 11 sts. 18 sts.

***Distribute these 18 sts over 3 needles.

Work straight until first finger measures 7cm/2¾in.

Next round [K2tog] 9 times. 9 sts.

Work 1 round.

Next round [K1, K2tog] 3 times.

Break yarn and thread through rem 6 sts. Pull up tight and fasten off securely.

Shape second finger

Return to sts left on holder and work as folls:

Next round K first 8 sts on holder, leave next 28 sts on holder, turn and cast on 2 sts, turn and K rem 8 sts from holder, then pick up and K 2 sts from base of first finger. 20 sts.

Distribute these 20 sts over 3 needles.

Work straight until second finger measures 7.5cm/3in.

Next round [K2tog] 10 times. 10 sts.

Work 1 round.

Next round [K2tog] 5 times.

Break yarn and thread through rem 5 sts. Pull up tight and fasten off securely.

Shape third finger

Return to sts left on holder and work as folls:

Next round K first 7 sts on holder, leave next 14 sts on holder, turn and cast on 2 sts, turn and K rem 7 sts from holder, then pick up and K 2 sts from base of second finger. 18 sts.

Distribute these 18 sts over 3 needles.

Work straight until third finger measures 7cm/2¾in.

Next round [K2tog] 9 times. 9 sts.

Work 1 round.

Next round [K1, K2tog] 3 times.

Break yarn and thread through rem 6 sts. Pull up tight and fasten off securely.

Shape fourth finger

Return to sts left on holder and work as folls:

Next round K rem 14 sts on holder, then pick up and K 2 sts from base of third finger. 16 sts.

Distribute these 16 sts over 3 needles.

Work straight until fourth finger measures 6cm/2¼in.

Next round [K2tog] 8 times. 8 sts.

Work 1 round.

Next round [K2tog, K1] twice, K2tog.

Break yarn and thread through rem 5 sts. Pull up tight and fasten off securely.

Shape thumb

Return to sts left on safety pin and work as folls:

Next round K 18 sts from safety pin, then pick up and K 6 sts from base of palm. 24 sts.

Distribute these 24 sts over 3 needles.

Work straight until thumb measures 6cm/2¼in.

Next round [K2tog] 12 times. 12 sts.

Work 1 round.

Next round [K2tog] 6 times.

Break yarn and thread through rem 6 sts. Pull up tight and fasten off securely.

LEFT GLOVE

Work as given for Right Glove to **.

Shape thumb

Round 25 Patt 54 sts, slip rem 18 sts onto a safety pin, turn and cast on 6 sts, turn. 60 sts.

Keeping patt correct (by now repeating the 12-st patt rep 5 times around each round), work 13 rounds, ending after chart round 10.

Break off A, B and C and cont in rounds of st st (K every round) using MC only.

Work 4 rounds.

Shape first finger

Next round K11, slip next 44 sts onto a holder, turn and cast on 2 sts, turn and K rem 5 sts. 18 sts.

Complete as given for right glove from ***.

MAKING UP

Press lightly following instructions on yarn label and avoiding ribbing.

Felted slippers

CATHERINE TOUGH

Size
The finished slippers measure 27cm/10½in from heel
to toe.

Yarns
Rowan *Scottish Tweed DK*:
A	Storm Grey 004	1 x 50g/1¾oz balls
B	Grey Mist 001	1 x 50g/1¾oz balls
C	Rose 026	1 x 50g/1¾oz balls

Needles
Pair of 5mm (UK no 6) (US size 8) knitting needles

Extras
30cm/11¾in square piece of thin foam rubber and
matching sewing thread

Tension
Before felting: 17 sts and 26 rows to 10cm/4in
measured over st st using 5mm (US size 8) needles
or size to obtain correct tension.

Abbreviations
See page 93.

SOLE SECTIONS (make 2)
Using 5mm (US size 8) needles and A, cast on 60 sts.
Starting with a K row, work in st st for 75cm/29½in.
Cast off.

UPPER SECTIONS (make 2)
Using 5mm (US size 8) needles and B, cast on 70 sts.
Starting with a K row, work in st st for 15cm/6in,
ending with RS facing for next row.
Join in C.

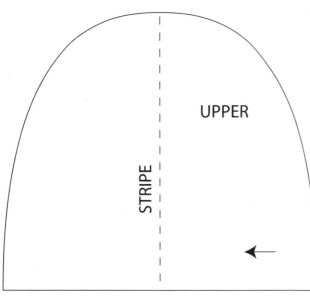

SOLE

TEMPLATES 50% OF ACTUAL SIZE

UPPER

STRIPE

← indicates direction of knitting

Using C, work in st st for 1 row.
Break off C.
Using B, and beg with a P row, cont in st st until
Section measures 15cm/6in from stripe in C.
Cast off.

MAKING UP

Machine wash knitted pieces at 40°/104°F to shrink
and felt them. The fabric will shrink a little more each
time it is washed so repeat this process until either
you are happy with the felted effect created, or
sections have shrunk to their minimum size. Felted
Sole piece must be at least 28cm/11in by 60cm/23½in,
and felted Upper piece must be at least 33cm/13in by
16cm/6¼in.

Once felted pieces are dry, press carefully from WS.
Enlarge Sole and Upper templates. Using template,
cut out Sole shape 4 times from felted sole fabric, and
twice from foam rubber. Trim away 6mm/¼in from
outer edge of foam rubber sole pieces.

Using template, cut out Upper shape twice from felted
upper fabric, ensuring stripe falls along line indicated.
Machine stitch twice along straight (row end) edge of
Upper – firstly 6mm/¼in from edge, then again
between this line of stitching and cut edge.

For Sole, form a "sandwich" by placing one foam
rubber piece against WS of one felted piece and
covering this with another felted sole piece, again with
WS of knitting against foam rubber. Tack pieces
together around entire outer edge.

Lay Upper onto Sole "sandwich", placing one end of
stripe at centre of toe and matching cut edges along
side – Upper will stand away from Sole to allow foot
to fit into slipper. Tack upper in place along outer
curved edge.

Machine stitch twice around entire outer edge of Sole,
enclosing Upper in stitching – firstly stitch 6mm/¼in
from edge, then again between this line of stitching
and cut edge.

Man's cabled sweater

WENDY BAKER

Sizes

	S-M	L-XL	XXL-XXXL	
To fit chest	97-102	107-112	117-122	cm
	38-40	42-44	46-48	in
Finished measurements				
Around chest	122	132	142	cm
	48	52	56	in
Length to shoulder	68	70	72	cm
	26³/₄	27¹/₂	28¹/₄	in
Sleeve seam	49	50	51	cm
	19¹/₄	19¹/₂	20	in

Yarns

10 (12: 13) x 100g/3¹/₂oz balls of Rowan *Scottish Tweed Chunky* in Olive Green 035.

Needles

Pair of 7mm (UK no 2) (US size 10¹/₂) knitting needles
Pair of 8mm (UK no 0) (US size 11) knitting needles
Cable needle

Tension

12 sts and 16 rows to 10cm/4in measured over st st using 8mm (US size 11) needles *or size to obtain correct tension.*

Abbreviations

See page 93.

Special abbreviations

wyaf = with yarn at front (RS) of work.
Cr4R = slip next st onto cable needle and leave at back of work, K3, then P1 from cable needle.
Cr4L = slip next 3 sts onto cable needle and leave at front of work, P1, then K3 from cable needle.
C6B = slip next 3 sts onto cable needle and leave at back of work, K3, then K3 from cable needle.
C6F = slip next 3 sts onto cable needle and leave at front of work, K3, then K3 from cable needle.

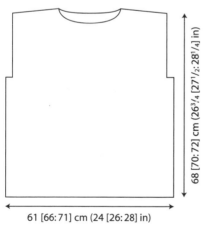

68 [70:72] cm (26³/₄ [27¹/₂:28¹/₄] in)

61 [66:71] cm (24 [26:28] in)

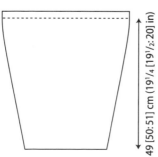

49 [50:51] cm (19¹/₄ [19¹/₂:20] in)

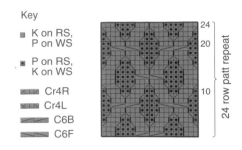

Key

- ■ K on RS, P on WS
- ▪ P on RS, K on WS
- ▨ Cr4R
- ▨ Cr4L
- ▨ C6B
- ▨ C6F

24 row patt repeat

BACK

Using 7mm (US size 10½) needles, cast on 73 (79: 85) sts.

Row 1 (WS) Purl.

Row 2 P2 (1: 4), *sl 5 wyaf, P3; rep from * to last 7 (6: 9) sts, sl 5 wyaf, P2 (1: 4).

Row 3 K2 (1: 4), *P5, K3; rep from * to last 7 (6: 9) sts, P5, K2 (1: 4).

Row 4 P2 (1: 4), *K2, inserting right needle point under strand of yarn of row 2 K next st, K2, P3; rep from * to last 7 (6: 9) sts, K2, inserting right needle point under strand of yarn of row 2 K next st, K2, P2 (1: 4).

These 4 rows form fancy rib.

Work in fancy rib for 13 rows more, ending with RS facing for next row.

Change to 8mm (US size 11) needles.

Starting with a P row, work in rev st st until Back measures 44 (45: 46)cm/17¼ (17¾: 18)in from cast-on edge, ending with RS facing for next row.

Shape armholes

Cast off 4 sts at beg of next 2 rows. 65 (71: 77) sts.

Work straight until armhole measures 23 (24: 25)cm/ 9 (9½: 9¾)in, ending with RS facing for next row.

Shape back neck

Next row (RS) P22 (24: 26) and turn, leaving rem sts on a holder.

Work each side of neck separately.

Dec 1 st at neck edge of next row, ending with RS facing for next row. 21 (23: 25) sts.

Shape shoulder

Cast off 10 (11: 12) sts at beg and dec 1 st at end of next row.

Work 1 row.

Cast off rem 10 (11: 12) sts.

With RS facing, rejoin yarn to rem sts, cast off centre 21 (23: 25) sts, P to end.

Complete to match first side, reversing shapings.

FRONT

Using 7mm (US size 10½) needles, cast on 73 (79: 85) sts.

Work in fancy rib as given for Back for 16 rows, ending with **WS** facing for next row.

Row 17 (WS) P30 (33: 36) sts, [M1, P2] 6 times, M1, P to end. 80 (86: 92) sts.

Change to 8mm (US size 11) needles.

Place cable panel

Row 1 (RS) P30 (33: 36), work next 20 sts as chart row 1 of cable panel, P to end.

Row 2 K30 (33: 36), work next 20 sts as chart row 2 of cable panel, K to end.

These 2 rows set position of cable panel with rev st st at sides.

Work straight in patt, repeating the 24 row patt rep throughout, until Front matches Back to start of armhole shaping, ending with RS facing for next row.

Shape armholes

Keeping patt correct, cast off 4 sts at beg of next 2 rows. 72 (78: 84) sts.

Work straight until 8 (8: 10) rows less have been worked than on Back to start of shoulder shaping, ending with RS facing for next row.

Shape neck

Next row (RS) P25 (27: 30) and turn, leaving rem sts on a holder.

Work each side of neck separately.

Dec 1 st at neck edge of next 4 rows, then on foll 1 (1: 2) alt rows. 20 (22: 24) sts.

Work 1 row, ending with RS facing for next row.

Shape shoulder

Cast off 10 (11: 12) sts at beg of next row.

Work 1 row.

Cast off rem 10 (11: 12) sts.

With RS facing, rejoin yarn to rem sts, cast off centre 22 (24: 24) sts, P to end.

Complete to match first side, reversing shapings.

SLEEVES

Using 7mm (US size 10½) needles, cast on 33 (35: 37) sts.

Row 1 (WS) Purl.

Row 2 P2 (3: 4), *sl 5 wyaf, P3; rep from * to last 7 (8: 9) sts, sl 5 wyaf, P2 (3: 4).

Row 3 K2 (3: 4), *P5, K3; rep from * to last 7 (8: 9) sts, P5, K2 (3: 4).

Row 4 P2 (3: 4), *K2, inserting right needle point under strand of yarn of row 2 K next st, K2, P3; rep from * to last 7 (8: 9) sts, K2, inserting right needle point under strand of yarn of row 2 K next st, K2, P2 (3: 4).

These 4 rows form fancy rib.

Work in fancy rib for 8 rows more, ending with WS facing for next row.

Row 13 (WS) Patt 11 (12: 13) sts, [M1, patt 2 sts] 6 times, M1, patt to end. 40 (42: 44) sts.

Change to 8mm (US size 11) needles.

Place cable panel

Row 1 (RS) P10 (11: 12), work next 20 sts as chart row 1 of cable panel, P to end.

Row 2 K10 (11: 12), work next 20 sts as chart row 2 of cable panel, K to end.

These 2 rows set position of cable panel with rev st st at sides.

Cont as set, inc 1 st at each end of next and every foll 4th row to 54 (60: 66) sts, then on every foll 6th row until there are 64 (68: 72) sts, taking inc sts into rev st st.

Work straight until Sleeve measures 49 (50: 51)cm/ 19¼ (19½: 20)in from cast-on edge, ending with RS facing for next row.

Shape top

Place markers at both ends of last row.

Work 6 rows more, ending with RS facing for next row.

Cast off.

MAKING UP

Press lightly on WS following instructions on yarn label and avoiding ribbing and cables.

Sew right shoulder seam.

Collar

With RS facing and using 7mm (US size 10½) needles, pick up and knit 8 (9: 12) sts down left side of neck, 15 (18: 18) sts from Front, 8 (9: 12) sts up right side of neck, then 34 (37: 39) sts from Back. 65 (73: 81) sts.

Starting with a P row, work in st st until Collar measures 7cm/2¾in from pick-up row, ending with RS of body facing for next row.

Now work in fancy rib as folls:

Row 1 (WS of Collar, RS of body) Purl.

Row 2 P2, *sl 5 wyaf, P3; rep from * to last 7 sts, sl 5 wyaf, P2.

Row 3 K2, *P5, K3; rep from * to last 7 sts, P5, K2.

Row 4 P2, *K2, inserting right needle point under strand of yarn of row 2 K next st, K2, P3; rep from * to last 7 sts, K2, inserting right needle point under strand of yarn of row 2 K next st, K2, P2.

These 4 rows form fancy rib.

Work in fancy rib for 16 rows more, ending with **WS** of Collar facing for next row.

Cast off in patt (on **WS**).

Sew left shoulder and Collar seam, reversing collar seam for turn-back. Matching sleeve markers to top of side seams and centre of sleeve cast-off edge to shoulder seam, sew Sleeves into armholes. Sew side and sleeve seams.

Fair Isle stripe cushion

SARAH DALLAS

Size
The finished cushion cover fits a 46cm/18in square cushion pad.

Yarns
5 x 50g/1¾oz balls of *Rowan Scottish Tweed DK* in main colour **MC** (Storm Grey 004) and one ball each in **A** (Brilliant Pink 010), **B** (Lavender 005), **C** (Porridge 024) and **D** (Midnight 023)

Needles
Pair of 3¾mm (UK no 9) (US size 5) knitting needles
Pair of 4½mm (UK no 7) (US size 7) knitting needles

Extras
46cm/18in square cushion pad

Tension
20 sts and 26 rows to 10cm/4in measured over st st using 4½mm (US size 7) needles *or size to obtain correct tension*.

Abbreviations
See page 93.

Special note
When working patt in st st from chart, strand yarn not in use loosely across WS of work, weaving it in every 3 or 4 sts. Work odd-numbered rows as K rows, reading them from right to left, and even-numbered rows as P rows, reading them from left to right.

CUSHION COVER
Using 3¾mm (US size 5) needles and A, cast on 91 sts.
Break off A and join in B.
Row 1 (RS) P1, *K1, P1; rep from * to end.
Row 2 K1, *P1, K1; rep from * to end.
These 2 rows form rib.
Work in rib for 4 rows more, ending with RS facing for next row.
Break off B and join in MC.
Change to 4½mm (US size 7) needles.
Beg with a K row, work in st st until Cover measures 31cm/12¼in from cast-on edge.
Place markers at both ends of last row.
Work straight until Cover measures 12cm/4¾in from markers, ending with RS facing for next row.
Work first Fair Isle band
Join in B.
Rows 1 and 2 Using B, knit.
Joining in and breaking off C and D as required, work 9 rows from chart, ending with **WS** facing for next row.
Rows 12 and 13 Using B, purl.
Break off B.
Beg with a P row and using MC only, work straight in st st until Cover measures 14cm/5½in from last row worked using B, ending with RS facing for next row.

Key
□ C
■ D

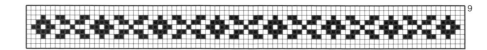

Work second Fair Isle band

Join in A.

Rows 1 and 2 Using A, knit.

Joining in and breaking off C and D as required, work 9 rows from chart, ending with **WS** facing for next row.

Rows 12 and 13 Using A, purl.

Break off A.

Beg with a P row and using MC only, work straight in st st until Cover measures 46cm/18in from markers, ending with RS facing for next row.

Place second set of markers at both ends of last row.

Work straight until Cover measures 26cm/10¼in from second set of markers, ending with **WS** facing for next row.

Break off MC and join in A.

Next row (WS) Purl.

Change to 3¾mm (US size 5) needles.

Work in rib for 5 rows, ending with **WS** facing for next row.

Break off A and join in B.

Next row (RS) Knit.

Cast off purlwise.

MAKING UP

Press lightly on WS following instructions on yarn label and avoiding ribbing.

With RS together, fold Cover level with first set of markers, then fold again level with second set of markers – cast-on and cast-off edges will overlap by about 13cm/5in. Sew row-end edges together to form side seams. Turn RS out and insert cushion pad.

Chunky stripe cushion

SARAH DALLAS

Size
The finished cushion cover fits a 46cm/18in square cushion pad.

Yarns
3 x 100g/3½oz balls of Rowan *Scottish Tweed Chunky* in main colour **MC** (Lewis Grey 007) and one ball each in **A** (Claret 013) and **B** (Midnight 023)

Needles
Pair of 7mm (UK no 2) (US size 10½) knitting needles
Pair of 8mm (UK no 0) (US size 11) knitting needles

Extras
46cm/18in square cushion pad

Tension
12 sts and 16 rows to 10cm/4in measured over st st using 8mm (US size 11) needles *or size to obtain correct tension.*

Abbreviations
See page 93.

Special note
When working patt in st st from chart, strand yarn not in use loosely across WS of work, weaving it in every 3 or 4 sts. Work odd-numbered rows as K rows, reading them from right to left, and even-numbered rows as P rows, reading them from left to right.

CUSHION COVER
Using 7mm (US size 10½) needles and A, cast on 53 sts.
Break off A and join in MC.
Row 1 (RS) P1, *K1, P1; rep from * to end.
Row 2 K1, *P1, K1; rep from * to end.
These 2 rows form rib.
Work in rib for 1 row more, ending with **WS** facing for next row.
Change to 8mm (US size 11) needles.
Beg with a P row, work in st st until Cover measures 31cm/12¼in from cast-on edge.
Place markers at both ends of last row.
Work straight until Cover measures 16cm/6¼in from markers, ending with RS facing for next row.
Work Fair Isle band
Join in A.
Rows 1 and 2 Using A, knit.
Break off A.
Joining in and breaking off B as required, work 19 rows from chart, ending with **WS** facing for next row.
Join in A.
Rows 22 and 23 Using A, purl.
Break off A.
Beg with a P row, work straight in st st using MC until

Key

▢ MC

■ B

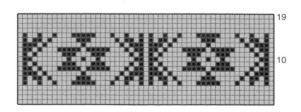

Cover measures 46cm/18in from markers, ending with RS facing for next row.

Place second set of markers at both ends of last row.

Break off MC and join in B.

Work straight until Cover measures 23cm/9in from second set of markers, ending with RS facing for next row.

Change to 7mm (US size 10½) needles.

Work in rib for 4 rows, ending with RS facing for next row.

Break off B and join in A.

Using A, cast off in rib.

MAKING UP

Press lightly on WS following instructions on yarn label and avoiding ribbing.

With RS together, fold Cover level with first set of markers, then fold again level with second set of markers – cast-on and cast-off edges will overlap by about 11cm/4¼in. Sew row-end edges together to form side seams. Turn RS out and insert cushion pad.

Patchwork throw

CATHERINE TOUGH

Size

The finished throw measures 95cm/37¹/₂in by 166cm/65¹/₄in.

Yarns

Rowan *Scottish Tweed DK*:

A	Grey Mist 001	8 x 50g/1³/₄oz balls
B	Storm Grey 004	4 x 50g/1³/₄oz balls
C	Purple Heather 030	2 x 50g/1³/₄oz balls
D	Autumn 029	2 x 50g/1³/₄oz balls
E	Herring 008	3 x 50g/1³/₄oz balls

Needles

Pair of 4¹/₂mm (UK no 7) (US size 7) knitting needles
Cable needle

Tension

20 sts and 26 rows to 10cm/4in measured over st st using 4¹/₂mm (US size 7) needles *or size to obtain correct tension.*

Abbreviations

See page 93.

Special abbreviation

C6B = slip next 3 sts onto cable needle and leave at back of work, K3, then K3 from cable needle.

FIRST SIDE STRIP

Using 4½mm (US size 7) needles and A, cast on 52 sts.
Row 1 (RS) [K4, P1] twice, [K6, P1, K4, P1] 3 times, K6.
Row 2 P6, [K1, P4, K1, P6] 3 times, [K1, P4] twice.
Rows 3 to 10 As rows 1 and 2, 4 times.
Row 11 [K4, P1] twice, [C6B, P1, K4, P1] 3 times, K6.
Row 12 As row 2.
Rows 13 to 20 As rows 1 and 2, four times.
These 20 rows form patt.
Work in patt until Strip measures 158cm/62¼in,
ending with RS facing for next row.
Cast off in patt.

CENTRE STRIP

Using 4½mm (US size 7) needles and B, cast on 77 sts.
Starting with a K row, work in st st for 25cm/9¾in,
ending with RS facing for next row.
Break off B and join in C.
Cont in st st until Strip measures 46cm/18in from
cast-on edge, ending with RS facing for next row.
Break off C and join in D.
Cont in st st until Strip measures 88cm/34½in from
cast-on edge, ending with RS facing for next row.
Break off D and join in E.
Cont in st st until Strip measures 126cm/49½in from
cast-on edge, ending with RS facing for next row.
Break off E and join in A.
Cont in st st until Strip measures 138cm/54¼n from
cast-on edge, ending with RS facing for next row.
Break off A and join in B.
Cont in st st until Strip measures 158cm/62¼in from
cast-on edge, ending with RS facing for next row.
Cast off.

SECOND SIDE STRIP

Using 4½mm (US size 7) needles and A, cast on 61 sts.
Row 1 (WS) Knit.
Row 2 *Insert right point needle into next st as though
to K it, wrap yarn round **both** needle points by taking it
under right needle point and over left needle point,
now wrap yarn around right needle point **only** in the
usual way as if to K a st, draw loop now on right needle
point through st and wrapped loop on left needle and
drop st and loop off left needle, rep from * to end.
These 2 rows form lace patt.
Work in lace patt until Strip measures 79cm/31in,
ending with RS facing for next row.
Break off A and join in C.
Starting with a K row, work in st st until Strip measures
101cm/39¾in from cast-on edge, ending with RS facing
for next row.
Break off C and join in B.
Next row Knit.
Starting with row 1, cont in lace patt until Strip
measures 158cm/62¼in from cast-on edge, ending with
RS facing for next row.
Cast off.

MAKING UP

Press lightly on WS following instructions on yarn label.
Using photograph as a guide and matching cast-on and
cast-off edges, sew Side Strips to either side of Centre
Strip.

Borders (both alike)

Using 4½mm (US size 7) needles and E, cast on 8 sts.
Work in garter st (K every row) until Border, when
slightly stretched, fits along entire cast-on (or cast-off)
edge of joined Strips, ending with RS facing for next row.
Cast off.
Sew Borders to cast-on and cast-off edges of joined
Strips as in photograph.

Child's jacket

CATHERINE TOUGH

Sizes

To fit age, approx	1-2	2-3	4-5	years
To fit chest	51	56	61	cm
	20	22	24	in
Finished measurements				
Around chest	61	67	72	cm
	24	26½	28½	in
Length to shoulder	30	34	39	cm
	11¾	13½	15½	in
Sleeve seam	20	26	30	cm
	7¾	10¼	11¾	in

Yarns

4 (5: 6) x 50g/1¾oz balls of Rowan *Scottish Tweed DK* in **MC** (Herring 008)

1 (1. 1) x 25g/⅞oz ball of Rowan *Scottish Tweed 4 ply* in **A** (Machair 002)

Needles

Pair of 3¼mm (UK no 10) (US size 3) knitting needles

Extras

3 buttons

Tension

21 sts and 36 rows to 10cm/4in measured over garter st using 3¼mm (US size 3) needles *or size to obtain correct tension*.

Abbreviations

See page 93.

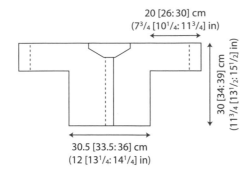

20 [26: 30] cm
(7³/₄ [10¹/₄: 11³/₄] in)

30 [34: 39] cm
(11³/₄ [13¹/₂: 15¹/₂] in)

30.5 [33.5: 36] cm
(12 [13¹/₄: 14¹/₄] in)

POCKET

Using 3¹/₄mm (US size 3) needles and A DOUBLE, cast on 7 sts.

Row 1 (RS) Inc in first st, K to last st, inc in last st.

Row 2 Purl.

Rep last 2 rows 4 (5: 6) times more, ending with RS facing for next row. 17 (19: 21) sts.

Starting with a K row, work in st st for 6 rows, ending with RS facing for next row.

Break off yarn and leave sts on a holder.

LEFT FRONT

Using 3¹/₄mm (US size 3) needles and A DOUBLE, cast on 35 (38: 41) sts.

Starting with a K row, work in st st for 4 rows, ending with RS facing for next row.

Break off A and join in MC SINGLE.

Work in garter st (K every row) until Left Front measures 20 (23: 27)cm/7³/₄ (9: 10¹/₂)in from first row using MC, ending with RS facing for next row.

Shape for sleeve

Next row (RS) Using MC cast on 48 (61: 70) sts, using A DOUBLE cast on 2 sts, using A K2, using MC K to end. 85 (101: 113) sts.

Twisting yarns together where they meet on WS to avoid holes forming, patt as folls:

Next row Using MC K to last 2 sts, using A DOUBLE K2.

Next row Using A DOUBLE K2, using MC K to end. These 2 rows form patt.

Cont in patt until Left Front measures 5 (6: 6)cm/ 2 (2¹/₄: 2¹/₄)in from sleeve cast-on edge, ending with

WS facing for next row.

Shape neck

Keeping patt correct, cast off 7 (8: 7) sts at beg of next row. 78 (93: 106) sts.

Next row (RS) Patt to last 4 sts, K2tog, K2.

Next row K2, K2tog, patt to end. 76 (91: 104) sts.

Working all neck decreases as set by last 2 rows, dec 1 st at neck edge of next 12 rows, then on foll 0 (0: 1) alt row. 64 (79: 91) sts.

Work straight until Left Front measures 10 (11: 12)cm/ 4 (4¹/₂ : 4³/₄)in from sleeve cast-on edge, ending with WS facing for next row.

Break off yarn and leave sts on a holder.

Mark positions for 3 buttons along left front opening edge – first to come level with row 25 of garter st section, last to come just below neck shaping and rem button evenly spaced between.

RIGHT FRONT

Using 3¹/₄mm (US size 3) needles and A DOUBLE, cast on 35 (38: 41) sts.

Starting with a K row, work in st st for 4 rows, ending with RS facing for next row.

Break off A and join in MC SINGLE.

Work in garter st (K every row) for 24 rows, ending with RS facing for next row.

Row 25 (RS) K2, K2tog, yfwd (to make a buttonhole), K to end.

Making a further 2 buttonholes in this way to correspond with positions marked for buttons on Left Front and noting that no further reference will be made to buttonholes, cont as folls:

Work 24 (26: 30) rows, ending with WS facing for next row.

Place pocket

Next row (WS) K9, cast off next 17 (19: 21) sts knitwise, K to end.

Next row K9 (10: 11), with RS facing K 17 (19: 21) sts from Pocket holder, K to end.

Work straight until Right Front measures 20 (23: 27)cm /7³/₄(9: 10¹/₂)in from first row using MC, ending with WS facing for next row.

Shape for sleeve

Next row (WS) Using MC cast on 48 (61: 70) sts, using A DOUBLE cast on 2 sts, using A K2, using MC K to end. 85 (101: 113) sts.

Twisting yarns together where they meet on WS to avoid holes forming, patt as folls:

Next row Using MC K to last 2 sts, using A DOUBLE K2.

Next row Using A DOUBLE K2, using MC K to end.

These 2 rows form patt.

Cont in patt until Right Front measures 5 (6: 6)cm/ 2 (2¼: 2¼)in from sleeve cast-on edge, ending with RS facing for next row.

Shape neck

Keeping patt correct, cast off 7 (8: 7) sts at beg of next row. 78 (93: 106) sts.

Working all neck decreases as set by Left Front, dec 1 st at neck edge of next 14 rows, then on foll 0 (0: 1) alt row. 64 (79: 91) sts.

Work straight until Right Front measures 10 (11: 12)cm/4 (4½: 4¾)in from sleeve cast-on edge, ending with **WS** facing for next row.

Do NOT break off yarn.

BACK

With **WS** facing, patt 64 (79: 91) sts of Right Front, turn and cast on 36 (38: 38) sts (for back neck), turn and with **WS** facing patt 64 (79: 91) sts of Left Front. 164 (196: 220) sts.

Work straight in patt as set (with 2 sts in garter st using A at **both** ends of rows) until Back measures 10 (11: 12)cm/4 (4½: 4¾)in from back neck cast-on edge, ending with RS facing for next row.

Shape sleeves

Cast off 50 (63: 72) sts at beg of next 2 rows. 64 (70: 76) sts.

Work straight in garter st using MC only until Back measures 20 (23: 27)cm/7¾ (9: 10½)in from Sleeve cast-off edge, ending with RS facing for next row.

Break off MC and join in A DOUBLE.

Starting with a K row, work in st st for 4 rows, ending with RS facing for next row.

Cast off.

MAKING UP

Press lightly on WS following instructions on yarn label.

Sew side and sleeve seams, reversing sleeve seam for first 5cm/2in for turn-back. Fold 4cm/1½in cuff to RS.

Sew on buttons.

Catch-stitch pocket lining on WS.

Fair Isle socks

SARAH DALLAS

Size
The finished socks measure 28cm/11in from heel to toe (*adjustable*), 18cm/7in around foot, and 28cm/11in around calf.

Yarns
Version 1 (with contrast trim):
3 x 25g/⁷⁄₈oz balls of Rowan *Scottish Tweed 4 ply* in main colour **MC** (Grey Mist 001) and one ball each in **A** (Midnight 023), **B** (Lewis Grey 007), **C** (Porridge 024) and **D** (Brilliant Pink 010)
Version 2 (without contrast trim):
3 x 25g/⁷⁄₈oz balls of Rowan *Scottish Tweed 4 ply* in main colour **MC** (Grey Mist 001) and one ball each in **A** (Midnight 023), **B** (Lewis Grey 007) and **C** (Porridge 024)

Needles
Set of four 3¼mm (UK no 10) (US size 3) double-pointed knitting needles

Tension
26 sts and 36 rounds to 10cm/4in measured over st st using 3¼mm (US size 3) needles *or size to obtain correct tension*.

Abbreviations
See page 93.

Special note
When working patt from chart, strand yarn not in use loosely across WS of work, weaving it in every 3 or 4 sts. Work ALL rounds as RS (knit) rounds, reading them from right to left.

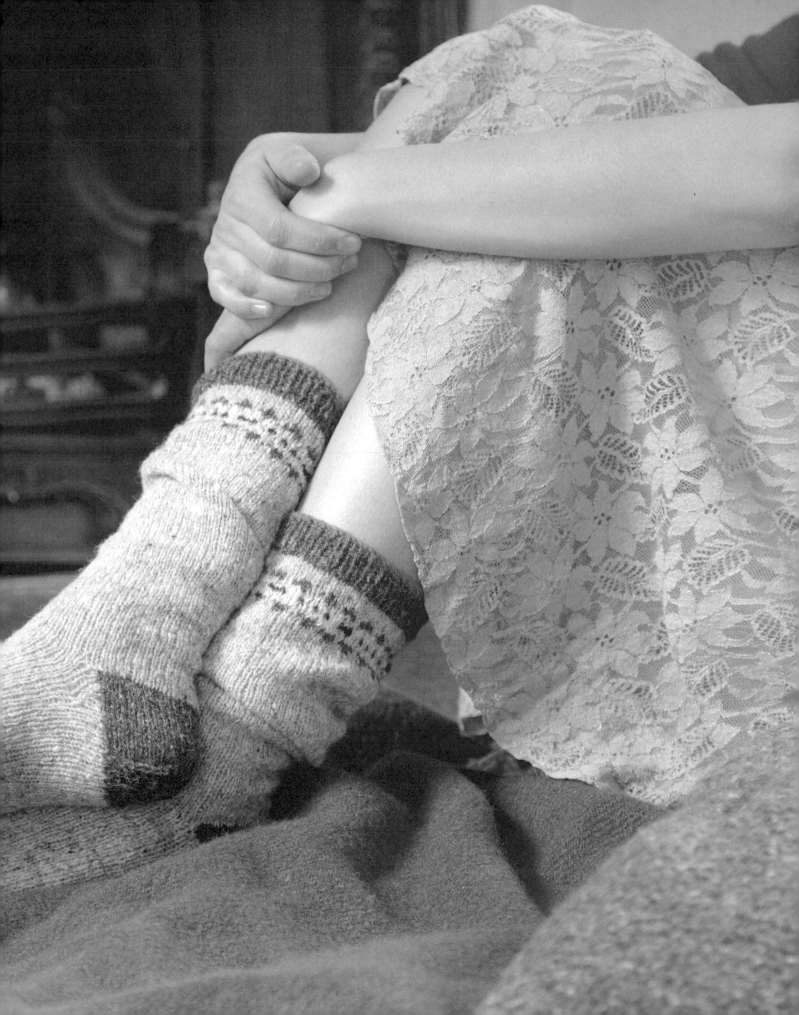

Key
□ MC
▦ B
□ C
■ A

12 st patt rep

14
10

SOCK (make 2)

Version 1 only

Using 3¼mm (US size 3) double-pointed needles and D, cast on 72 sts, distributing them evenly over 3 needles.

Break off D and join in B.

Round 1 (RS) *K1, P1; rep from * to end.

This round forms rib.

Work in rib for 9 rounds more, ending with RS facing for next row.

Break off B and join in MC.

Version 2 only

Using 3¼mm (US size 3) double-pointed needles and A, cast on 72 sts, distributing them evenly over 3 needles.

Break off A and join in MC.

Round 1 (RS) *K1, P1; rep from * to end.

This round forms rib.

Work in rib for 9 rounds more, ending with RS facing for next row.

Both versions

Joining in A, B and C as required, repeating the 12 st patt rep 6 times around each round, work from chart as folls:

Work rounds 1-14 of chart.

Break off contrasts and cont using MC only.

Round 15 (RS) Knit.

This round forms st st.

Work in st st for 19 rounds more.

Round 35 K2, K2tog, K to last 4 sts, skpo, K2. 70 sts.

Work 4 rounds.

Rep last 4 rounds 8 times more, then round 35 again. 52 sts.

Work straight until Sock measures 28cm/11in from cast-on edge.

Break off yarn.

Shape heel

Slip first and last 13 sts of next round onto a holder, leaving centre 26 sts on needle.

Join in B and now working in **rows**, not rounds, work heel as folls:

Starting with a K row, work in st st for 20 rows, ending with RS facing for next row.

Next row (RS) K17, skpo and turn.

Next row sl 1, P8, P2tog and turn.

Next row sl 1, K8, skpo and turn.

Rep last 2 rows 6 times more.

Next row sl 1, P8, P2tog. 10 sts.

Break off B.

With RS facing and using MC, with first needle pick up and knit 10 sts up first row-end edge of heel, and K 10 heel sts, with second needle pick up and knit 10 sts down second row-end edge of heel, then first 8 sts from holder, with third needle K rem 18 sts from holder. 56 sts.

Now working again in rounds of st st, cont as folls:

Next round (RS) K1, skpo, K24, K2tog, K to end. 54 sts.

Work 1 round.

Next round (RS) K1, skpo, K22, K2tog, K to end. 52 sts.

Work 1 round.

Next round (RS) K1, skpo, K20, K2tog, K to end. 50 sts.

Work 1 round.

Next round (RS) K1, skpo, K18, K2tog, K to end. 48 sts.

Work straight in rounds of st st until Sock measures 18cm/7in (*or required length*) from heel pick-up row.

Shape toe

Break off MC and join in B.

Round 1 and every foll alt round Knit.

Round 2 [K1, skpo, K18, K2tog, K1] twice. 44 sts.

Round 4 [K1, skpo, K16, K2tog, K1] twice. 40 sts.

Round 6 [K1, skpo, K14, K2tog, K1] twice. 36 sts.

Round 8 [K1, skpo, K12, K2tog, K1] twice. 32 sts.

Round 10 [K1, skpo, K10, K2tog, K1] twice. 28 sts.

Round 11 As round 1.

Cast off.

Making up

Press lightly on WS following instructions on yarn label and avoiding ribbing.

Sew toe seam.

Cabled blanket coat

SARAH DALLAS

Sizes

	XS-S	M-L	XL-XXL	
To fit bust	81-86	91-97	102-107	cm
	32-34	36-38	40-42	in
Finished measurements				
Width all round	152	161	170	cm
	59¾	63¼	67	in
Length to shoulder	52	54	56	cm
	20½	21¼	22	in
Sleeve seam	43	44	45	cm
	17	17¼	17¾	in

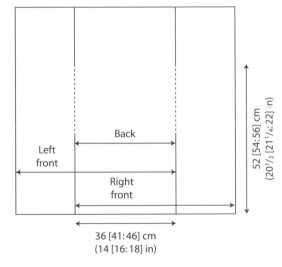

Back

Left front

Right front

52 [54:56] cm (20½ [21¼:22] in)

36 [41:46] cm (14 [16:18] in)

43 [44:45] cm (17 [17¼:17¾] in)

Yarns
18 (19: 20) x 50g/1¾oz balls of Rowan *Scottish Tweed DK* in main colour **MC** (Celtic Mix 022) and one ball in **A** (Midnight 023)

Needles
Pair of 3¾mm (UK no 9) (US size 5) knitting needles
Pair of 4½mm (UK no 7) (US size 7) knitting needles
Cable needle

Tension
21½ sts and 30 rows to 10cm/4in measured over patt using 4½mm (US size 7) needles *or size to obtain correct tension.*

Abbreviations
See page 93.

Special abbreviation
C8B = slip next 4 sts onto cable needle and leave at back of work, K4, then K4 from cable needle.

BODY
The body is worked in one piece, starting at right front opening edge.
Using 3¾mm (US size 5) needles and A, cast on 162 (166: 170) sts.
Break off A and join in MC.
Row 1 (RS) K2, *P2, K2; rep from * to end.
Row 2 P2, *K2, P2; rep from * to end.
These 2 rows form rib.
Work in rib for 9 rows more, ending with **WS** facing for next row.
Row 12 (WS) P2, K2, P2, K2tog, rib to end.
161 (165: 169) sts.

Change to 4½mm (US size 7) needles.

Now work in patt as folls:

Row 1 (RS) [K2, P2] 7 (8: 9) times, K2, *[P1, K1] twice, P2, K8, P1; rep from * to last 11 sts, [P1, K1] twice, P1, K2, P2, K2.

Row 2 P2, K2, P2, *[K1, P1] twice, K2, P8, K1; rep from * to last 35 (39: 43) sts, [K1, P1] twice, K1, [P2, K2] 7 (8: 9) times, P2.

Rows 3 to 6 As rows 1 and 2, twice.

Row 7 [K2, P2] 7 (8: 9) times, K2, *[P1, K1] twice, P2, C8B, P1; rep from * to last 11 sts, [P1, K1] twice, P1, K2, P2, K2.

Row 8 As row 2.

Rows 9 to 12 As rows 1 and 2, twice.

These 12 rows form patt.

Cont in patt until Body measures 58 (60: 62)cm/22¾ (23½: 24½)in from cast-on edge, ending with RS facing for next row.

**Shape armhole

Next row (RS) Patt 49 sts placing red marker on last of these sts, cast off next 56 (58: 60) sts, patt to end placing blue marker on first of these sts.

Next row Patt 56 (58: 60) sts, turn and cast on 56 (58: 60) sts, turn and patt to end. 161 (165: 169) sts.**

Work straight until Body measures 36 (41: 46)cm/14 (16: 18)in from armhole, ending with RS facing for next row.

Rep from ** to ** once more.

Work straight until Body measures 53 (55: 57)cm/20¾ (21½: 22½)in from second armhole, ending with **WS** facing for next row.

Next row (WS) P2, K2, P2, inc in next st, rib to end. 162 (166: 170) sts.

Work in rib as given for cast-on edge for 12 rows, ending with RS facing for next row.

Break off MC and join in A.

Cast off.

SLEEVES

Using 3¾mm (US size 5) needles and A, cast on 50 (54: 54) sts.

Break off A and join in MC.

Work in rib as given for Body for 7cm/2¾in, ending with RS facing for next row.

Change to 4½mm (US size 7) needles.

Cont in rib, shaping sides by inc 1 st at each end of next and every foll 6th row to 70 (72: 84) sts, then on every foll 8th row until there are 80 (84: 88) sts, taking inc sts into rib.

Work straight until Sleeve measures 43 (44: 45)cm/17 (17¼: 17¾)in from cast-on edge, ending with RS facing for next row.

Shape top

Keeping rib correct, cast off 3 (4: 5) sts at beg of next 2 rows. 74 (76: 78) sts.

Dec 1 st at each end of next 7 rows, then on every foll alt row until 22 sts rem.

Work 1 row, ending with RS facing for next row.

Cast off rem 22 sts.

MAKING UP

Press lightly on WS following instructions on yarn label and avoiding ribbing.

Sew sleeve seams. Matching top of sleeve seam to blue (underarm) marker and centre of sleeve cast-off edge to red (shoulder) marker, sew Sleeves into armholes.

Textured scarf

WENDY BAKER

Size

The finished scarf measures 28cm/11in by
141cm/55½in.

Yarns

4 x 25g/⅞oz balls of Rowan *Scottish Tweed 4 ply* in
main colour **MC** (Celtic Mix 022) and 2 balls each in
A (Herring 008) and **B** (Winter Navy 021)

Needles

Pair of 3¼mm (UK no 10) (US size 3) knitting needles

Tension

28 sts and 41 rows to 10cm/4in measured over patt
using 3¼mm (US size 3) needles *or size to obtain
correct tension.*

Abbreviations

See page 93.

SCARF

Using 3¼mm (US size 3) needles and MC, cast on
79 sts.

Row 1 (RS) K1, *P1, K1; rep from * to end.

Row 2 As row 1.

These 2 rows form moss st.

Work in moss st for 12 rows more, ending with RS
facing for next row.

Using a separate ball of yarn for each block of colour
and twisting yarns together on WS where they meet to
avoid holes forming, work in diamond patt as folls:

Row 1 (RS) Using MC moss st 9 sts, using A K1, [P1,
K9, P1, K1] 5 times, using MC moss st 9 sts.

Row 2 Using MC moss st 9 sts, using A K1, [P1, K1, P7,
K1, P1, K1] 5 times, using MC moss st 9 sts.

Row 3 Using MC moss st 9 sts, K1, *P1, K1, P1, K5,
[P1, K1] twice; rep from * 4 times more, moss st 9 sts.

Row 4 Using MC moss st 9 sts, P1, *[P1, K1] twice, P3,
K1, P1, K1, P2; rep from * 4 times more, moss st 9 sts.

Row 5 Using MC moss st 9 sts, using A K1, *K2, [P1,
K1] 3 times, P1, K3; rep from * 4 times more, using MC
moss st 9 sts.

Row 6 Using MC moss st 9 sts, using A P1, *P3, [K1,
P1] twice, K1, P4; rep from * 4 times more, using MC
moss st 9 sts.

Row 7 Using MC moss st 9 sts, K1, [K4, P1, K1, P1, K5]
5 times, moss st 9 sts.

Row 8 Using MC moss st 9 sts, P1, *P3, [K1, P1] twice,
K1, P4; rep from * 4 times more, moss st 9 sts.

Row 9 As row 5.

Row 10 Using MC moss st 9 sts, using A P1, *[P1, K1]
twice, P3, K1, P1, K1, P2; rep from * 4 times more,
using MC moss st 9 sts.

Row 11 As row 3.

Row 12 Using MC moss st 9 sts, K1, [P1, K1, P7, K1,
P1, K1] 5 times, moss st 9 sts.

These 12 rows form diamond patt and stripe sequence.
Cont as set until Scarf measures 26cm/10¼in from
cast-on edge, ending with RS facing for next row.
Break off A.

Work in diamond patt using MC **only** until Scarf
measures 48cm/18¾in from cast-on edge, ending with
WS facing for next row.

Next row (WS) Patt 9 sts, [inc in next st, patt 19 sts]
3 times, inc in next st, patt to end. 83 sts.

Using a separate ball of yarn for each block of colour
and twisting yarns together on WS where they meet to
avoid holes forming, now work in block patt as folls:

Row 1 (RS) Using MC moss st 9 sts, using B K65, using
MC moss st 9 sts.

Row 2 Using MC moss st 9 sts, using B K5, [P5, K5] 12 times, using MC moss st 9 sts.

Rows 3 to 6 As rows 1 and 2, twice.

Row 7 Using MC moss st 9 sts, using B K65, using MC moss st 9 sts.

Row 8 Using MC moss st 9 sts, using B P5, [K5, P5] 12 times, using MC moss st 9 sts.

Rows 9 to 12 As rows 7 and 8, twice.

These 12 rows form block patt.

Work in block patt until Scarf measures 93cm/36½in from cast-on edge, dec 1 st at centre of last row and ending with RS facing for next row. 82 sts.

Using a separate ball of yarn for each block of colour and twisting yarns together on WS where they meet to avoid holes forming, now work in double moss st patt as folls:

Row 1 (RS) Using MC moss st 9 sts, using A [K2, P2] 16 times, using MC moss st 9 sts.

Row 2 As row 1.

Row 3 Using MC moss st 9 sts, using A [P2, K2] 16 times, using MC moss st 9 sts.

Row 4 As row 3.

These 4 rows form double moss st patt.

Work in double moss st patt until Scarf measures approx 116cm/45½in from cast-on edge, ending after patt row 4 and with RS facing for next row.

Keeping double moss st patt correct, using a separate ball of yarn for each block of colour and twisting yarns together on WS where they meet to avoid holes forming, now work in double moss st patt in stripes as folls:

Row 1 (RS) Using MC moss st 9 sts, [K2, P2] 16 times, moss st 9 sts.

Row 2 As row 1.

Row 3 Using MC moss st 9 sts, using A [P2, K2] 16 times, using MC moss st 9 sts.

Row 4 As row 3.

These 4 rows form striped double moss st patt. Work in striped double moss st patt until Scarf measures approx 138cm/54¼in from cast-on edge, ending after patt row 3 and with **WS** facing for next row.

Next row (WS) Using MC moss st 9 sts, using A patt 9 sts, [work 2 tog, patt 20 sts] twice, work 2 tog, patt 9 sts, using MC moss st 9 sts. 79 sts.

Break off A and cont using MC **only**.

Work in moss st for 14 rows, ending with RS facing for next row.

Cast off in moss st.

MAKING UP

Press lightly on WS following instructions on yarn label.

Tie shrug

SARAH DALLAS

Sizes

	XS-M	L-XXL	
To fit bust	81-91	97-107	cm
	32-36	38-42	in
Finished measurements			
Width, excluding ties	100	110	cm
	39¼	43¼	in
Length	39	39	cm
	15¼	15¼	in

Yarn

9 (11) x 25g/⁷/₈oz balls of Rowan *Scottish Tweed 4 ply* in Heath 014

Needles

Pair of 5mm (UK no 6) (US size 8) knitting needles
Pair of 9mm (UK no 00) (US size 13) knitting needles

Tension

11 sts and 28 rows to 10cm/4in measured over patt using yarn DOUBLE and 9mm (US size 13) needles *or size to obtain correct tension.*

Abbreviations

See page 93.

Special abbreviation

K1 below = K into next st 1 row below and at same time slip off st above.

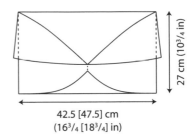

42.5 [47.5] cm
(16¾ [18¾] in)

27 cm (10¾ in)

MAIN SECTION

Using 9mm (US size 13) needles and yarn DOUBLE, cast on 93 (105) sts loosely.

Row 1 (RS) K2, *P1, K1; rep from * to last st, K1.

Now work in patt as folls:

Row 2 (WS) Sl 1, *P1, K1 below; rep from * to last 2 sts, P1, K1.

Row 3 Sl 1, *K1 below, P1; rep from * to last 2 sts, K1 below, K1.

Rows 2 and 3 form patt.

Work in patt for 20 rows more.

Change to 5mm (US size 8) needles.

Next row (WS) K1, *P1, K1; rep from * to end.

Next row K1, *K1, P1; rep from * to last 2 sts, K2.

Last 2 rows form rib.

Work in rib for 2 rows more.

Change to 9mm (US size 13) needles.

Work in patt for 20 rows.

Change to 5mm (US size 8) needles.

Work in rib for 4 rows.

Change to 9mm (US size 13) needles.

Work in patt for 18 rows.

Change to 5mm (US size 8) needles.

Work in rib for 4 rows.

Change to 9mm (US size 13) needles.

Work in patt for 16 rows.

Change to 5mm (US size 8) needles.

Work in rib for 4 rows.
Change to 9mm (US size 13) needles.
Work in patt for 12 rows.
Change to 5mm (US size 8) needles.
Work in rib for 8 rows, ending with **WS** facing for next row.
Cast off in rib.

END SECTIONS AND TIES (both alike)
With RS facing, using 9mm (US size 13) needles and yarn DOUBLE, pick up and knit 58 sts along one row-end edge of Main Section.
Row 1 (WS) [P2tog] 29 times. 29 sts.
Place marker on centre st of last row.
Row 2 K to within 1 st of marked st, sl 1, K2tog (marked st is first of these 2 sts), psso, K to end. 27 sts.
Row 3 K4, P to last 4 sts, K4.
Rep rows 2 and 3, nine times more. 9 sts.
Shape tie
Next row (RS) K2, yfwd, skpo, K1, K2tog, yfwd, K2.
Next row K2, P2tog, yfrn, P1, yon, K2tog, K1.
Rep last 2 rows until Tie measures 28cm/11in, ending with RS facing for next row.
Cast off.

MAKING UP
Press lightly on WS following instructions on yarn label.

Rever jacket

WENDY BAKER

Sizes

	S-M	L-XL	XXL-XXXL	
To fit chest	97-102	107-112	117-122	cm
	38-40	42-44	46-48	in

Finished measurements

	S-M	L-XL	XXL-XXXL	
Around chest	110	120	130	cm
	43¼	47¼	51	in
Length to shoulder	74	76	78	cm
	29	30	30¾	in
Sleeve seam	54	55	56	cm
	21¼	21½	22	in

Yarns

8 (9: 10) x 100g/3½oz balls of Rowan *Scottish Tweed Aran* in main colour **MC** (Lovat 033) and one ball in **A** (Machair 002)

Needles

Pair of 5½mm (UK no 5) (US size 9) knitting needles

Extras

4 buttons

Tension

16 sts and 23 rows to 10cm/4in measured over st st using 5½mm (US size 9) needles *or size to obtain correct tension*.

Abbreviations

See page 93.

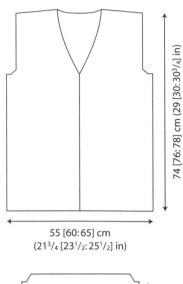

55 [60:65] cm
(21³/₄ [23¹/₂:25¹/₂] in)

74 [76:78] cm (29 [30:30³/₄] in)

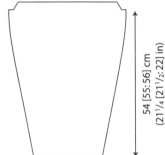

54 [55:56] cm
(21¹/₄ [21¹/₂:22] in)

BACK

Using 5¹/₂mm (US size 9) needles and MC, cast on 88 (96: 104) sts.

Starting with a K row, work in st st until Back measures 50 (51: 52)cm/19¹/₂ (20: 20¹/₂)in from cast-on edge, ending with RS facing for next row.

Shape armholes

Cast off 3 (4: 5) sts at beg of next 2 rows. 82 (88: 94) sts.

Next row (RS) K1, skpo, K to last 3 sts, K2tog, K1. 80 (86: 92) sts.

Working all armhole decreases as set by last row, dec 1 st at each end of 2nd and foll 2 (3: 4) alt rows. 74 (78: 82) sts.

Work straight until armhole measures 23 (24: 25)cm/ 9 (9¹/₂: 9³/₄)in, ending with RS facing for next row.

Shape back neck

Next row (RS) K25 (26: 27) and turn, leaving rem sts on a holder.

Work each side of neck separately.

Dec 1 st at neck edge of next row, ending with RS facing for next row. 24 (25: 26) sts.

Shape shoulder

Cast off 11 (12: 12) sts at beg and dec 1 st at end of next row.

Work 1 row.

Cast off rem 12 (12: 13) sts.

With RS facing, rejoin yarn to rem sts, cast off centre 24 (26: 28) sts, K to end.

Complete to match first side, reversing shapings.

POCKET LININGS (make 2)

Using 5¹/₂mm (US size 9) needles and A, cast on 28 sts.

Starting with a K row, work in st st for 40 rows, ending with RS facing for next row.

Break off yarn and leave sts on a holder.

RIGHT FRONT

Using 5¹/₂mm (US size 9) needles and MC, cast on 40 (44: 48) sts.

Starting with a K row, work in st st for 4 rows, ending with RS facing for next row.

Row 5 (RS) Cast on 8 sts, work across these 8 sts as folls: [K1, P1] 4 times, K to end. 48 (52: 56) sts.

Row 6 P to last 8 sts, [P1, K1] 4 times.

Row 7 [K1, P1] 4 times, K to end.

Rows 6 and 7 set the sts – front opening edge 8 sts in moss st with all other sts in st st.

Work as set for 37 rows more, ending with RS facing for next row.

Place pocket

Row 45 (RS) [K1, P1] 4 times, K6 (8: 10), slip next 28 sts onto a holder and, in their place, K across 28 sts of first Pocket Lining, K6 (8: 10).

Work straight until Right Front matches Back to start of armhole shaping, ending with RS facing for next row.

Shape collar

Next row (RS) Inc in first st (for collar increase), moss st 7 sts, place marker on needle, K to end. 49 (53: 57) sts.

Working all collar increases as set by last row and taking collar inc sts into moss st, cont as folls:

Shape armhole

Cast off 3 (4: 5) sts at beg of next row. 46 (49: 52) sts.

Shape front slope

Next row (RS) Moss st to marker, slip marker onto right needle, skpo (for front slope decrease), K to last 3 sts, K2tog (for armhole decrease), K1. 44 (47: 50) sts.

Working all armhole and front slope decreases and collar increases as now set, cont as folls:

Dec 1 st at armhole edge of 2nd and foll 2 (3: 4) alt rows **and at same time** dec 1 st at front slope edge of 4th and foll 0 (4th: 4th) row **and at same time** inc 1 st at collar edge of 4th and foll 0 (0: 6th) row. 41 (42: 45) sts.

Now keeping armhole edge straight, dec 1 st at front slope edge of 2nd (4th: 2nd) and 7 (7: 8) foll 4th rows **and at same time** inc 1 st at collar edge of 4th (2nd: 6th) and 5 (5: 4) foll 6th rows. 39 (40: 41) sts.

Work 5 (5: 1) rows, ending with RS facing for next row.

Shape rever

Next row (RS) Cast off 7 sts, cast on and moss st 9 sts, patt to end. 41 (42: 43) sts.

Work straight until Right Front matches Back to start of **shoulder** shaping (this is 2 rows **after** start of back neck shaping), ending with RS facing for next row.

Work 1 row, ending with **WS** facing for next row.

Shape shoulder

Cast off 11 (12: 12) sts at beg of next row, then 12 (12: 13) sts at beg of foll alt row. 18 sts.

Shape back collar extension

Next row (RS of Front, WS of Collar) Moss st 12 sts and turn.

Next row Sl 1, moss st to end.

Work in moss st across all sts for 6 rows.

Rep last 8 rows until shorter row-end edge of collar extension measures 9 (9.5: 10)cm/3½ (3¾: 4)in.

Cast off in moss st.

Mark positions for 4 buttons along right front opening edge – first to come level with row 11, last to come just below start of front slope shaping and rem 2 buttons evenly spaced between.

LEFT FRONT

Using 5½mm (US size 9) needles and MC, cast on 40 (44: 48) sts.

Starting with a K row, work in st st for 5 rows, ending with **WS** facing for next row.

Row 6 (WS) Cast on 8 sts, work across these 8 sts as folls: [K1, P1] 4 times, P to end. 48 (52: 56) sts.

Row 7 K to last 8 sts, [P1, K1] 4 times.

Row 8 [K1, P1] 4 times, P to end.

Rows 7 and 8 set the sts – front opening edge 8 sts in moss st with all other sts in st st.

Work as set for 2 rows more, ending with RS facing for next row.

Row 11 (RS) Patt to last 6 sts, work 2 tog, [yrn] twice, work 2 tog (to make a buttonhole), moss st 2 sts.

Row 12 Moss st 3 sts, [P1, K1] into double yrn of previous row, patt to end.

Working 3 more buttonholes as set by last 2 rows to correspond with positions marked for buttons on Right Front and noting that no further reference will be made to buttonholes, cont as folls:

Work 32 rows more, ending with RS facing for next row.

Place pocket

Row 45 (RS) K6 (8: 10), slip next 28 sts onto a holder and, in their place, K across 28 sts of second Pocket Lining, K6 (8: 10), [P1, K1] 4 times.

Work straight until Left Front matches Back to start of armhole shaping, ending with RS facing for next row.

Shape collar and armhole

Next row (RS) Cast off 3 (4: 5) sts, K to last 8 sts, place marker on needle, moss st 7 sts, inc in last st (for collar increase). 46 (49: 52) sts.

Working all collar increases as set by last row and taking collar inc sts into moss st, cont as folls:

Work 1 row.

Shape front slope

Next row (RS) K1, skpo (for armhole decrease), K to within 2 sts of marker, K2tog (for front slope decrease), slip marker onto right needle, moss st to end. 44 (47: 50) sts.

Working all armhole and front slope decreases and collar increases as now set, cont as folls:

Dec 1 st at armhole edge of 2nd and foll 2 (3: 4) alt rows **and at same time** dec 1 st at front slope edge of 4th and foll 0 (4th: 4th) row **and at same time** inc 1 st at collar edge of 4th and foll 0 (0: 6th) row. 41 (42: 45) sts.

Now keeping armhole edge straight, dec 1 st at front slope edge of 2nd (4th: 2nd) and 7 (7: 8) foll 4th rows **and at same time** inc 1 st at collar edge of 4th (2nd: 6th) and 5 (5: 4) foll 6th rows. 39 (40: 41) sts.

Work 4 (4: 0) rows, ending with **WS** facing for next row.

Shape rever

Next row (WS) Cast off 7 sts, cast on and moss st 9 sts, patt to end. 41 (42: 43) sts.

Work straight until Left Front matches Back to start of **shoulder** shaping (this is 2 rows **after** start of back neck shaping), ending with RS facing for next row.

Shape shoulder

Cast off 11 (12: 12) sts at beg of next row, then 12 (12: 13) sts at beg of foll alt row. 18 sts.

Shape back collar extension

Next row (WS of Front, RS of Collar) Moss st 12 sts and turn.

Next row Sl 1, moss st to end.

Work in moss st across all sts for 5 rows.

Rep last 8 rows until shorter row-end edge of collar extension measures 9 (9.5: 10)cm/3½ (3¾: 4)in.

Cast off in moss st.

SLEEVES

Using 5½mm (US size 9) needles and MC, cast on 40 (42: 44) sts.

Starting with a K row, work in st st for 16 rows, ending with RS facing for next row.

Row 17 (RS) K3, M1, K to last 3 sts, M1, K3. 42 (44: 46) sts.

Working all increases as set by last row, inc 1 st at each end of 6th and every foll 6th row until there are 66 (72: 80) sts.

S-M and L-XL sizes only

Inc 1 st at each end of every foll 8th row until there are 72 (76) sts.

All sizes

Work straight until Sleeve measures 54 (55: 56)cm/ 21¼ (21½: 22)in from cast-on edge, ending with RS facing for next row.

Shape top

Place markers at both ends of last row.

Work 4 (6: 8) rows more, ending with RS facing for next row.

Working all decreases in same way as for Back armhole decreases, dec 1 st at each end of next and foll 3 (4: 5) alt rows.

Work 1 row, ending with RS facing for next row.

Cast off rem 64 (66: 68) sts.

MAKING UP

Press lightly on WS following instructions on yarn label.

Sew shoulder seams. Sew centre back (cast-off edge) seam of collar extensions, then sew one edge to back neck. Matching sleeve markers to top of side seams and centre of sleeve cast-off edge to shoulder seam, sew Sleeves into armholes. Sew side and sleeve seams, reversing seams for final 4 rows of roll-back edging.

Pocket tops (both alike)

Slip 28 pocket sts onto 5½ mm (US size 9) needles and rejoin MC with RS facing.

Starting with a K row, work in st st for 2 rows, ending with RS facing for next row.

Cast off.

Sew Pocket Linings in place on inside, then neatly sew down ends of Pocket Tops. Sew on buttons.

USEFUL INFORMATION

The following notes will help you to knit the garments in this book successfully.

TENSION

Obtaining the correct tension is the factor that can make the difference between a garment that fits and one that does not. It controls both the shape and size of a knitted garment, so any variation, however slight, can distort the finished size. Different designers feature in our books and it is their tension, given at the start of each pattern, that you must match.

To check this against your own tension, we recommend that you knit a square in pattern and/or stocking stitch (depending on the pattern instructions) of perhaps 5–10 more stitches and 5–10 more rows than those given in the tension note. Mark out the central 10cm (4in) square with pins. If you have too many stitches to 10cm (4in) try again using larger sized needles, if you have too few stitches to 10cm (4in) try again using smaller sized needles.

Once you have achieved the correct tension your garment will be knitted to the measurements indicated in the size diagram shown with the pattern.

SIZING

The instructions in each pattern are given for the smallest size. The figures in brackets are for the larger sizes. Where there is one set of figures only it applies to all sizes.

All garment patterns include "ease" to allow for a comfortable fit. The bust measurement of the knitted garment at the start of each pattern includes this ease, and the actual measurements are shown in the size diagram that accompanies the garment pattern (the sketch of the finished garment and its dimensions). The size diagram shows the finished width of the garment at the under-arm point, and it is this measurement that you should use to choose an appropriate size. A useful tip is to measure one of your own garments that fits comfortably and choose a size that is similar. Having chosen a size based on width, look at the corresponding length for that size; if you are not happy with the total length that we recommend, adjust your own garment before beginning your armhole shaping – any adjustment after this point will mean that your sleeve will not fit into your garment easily; and don't forget to take your adjustment into account if there is any side seam shaping. Finally, look at the sleeve length; the size diagram shows the finished sleeve measurement, taking into account any top-arm insertion length. Measure your body between the centre of your neck and your wrist, this measurement should correspond to half the garment width plus the sleeve length. Again, your sleeve length may be adjusted, but remember to take into consideration your sleeve increases if you do adjust the length – you must increase more frequently than the pattern states to shorten your sleeve, less frequently to lengthen it.

CHART NOTE

Many of the patterns in the book are worked from charts. Each square on a chart represents a stitch and each line of squares represents a row of knitting. Each colour used is given a different letter and these are shown in the materials section, or in the key alongside the chart of each pattern. When working from the charts, read odd-numbered rows (K) from right to left and even-numbered rows (P) from left to right, unless otherwise stated.

KNITTING WITH MORE THAN ONE COLOUR

There are two main methods of working colour into a knitted fabric: intarsia and Fair Isle techniques. The first method produces a single thickness of fabric and is usually used where a colour is only required in a particular area of a row. Where a repeating pattern is created across the row, the Fair Isle technique is normally used.

Intarsia technique: The simplest method is to cut short lengths of yarn for each motif or block of colour used in a row. Then join in the various colours at the appropriate point on the row, linking one colour to the next by twisting them around each other where they meet on the wrong side to avoid gaps. All ends can then either be darned along the colour join lines, as each motif is completed or then can be " knitted-in" to the fabric of the knitting as each colour is worked into the pattern. This is done in much the same way as "weaving- in" yarns when working the Fair Isle technique and does save time darning-in ends. It is essential that the tension is noted for Intarsia as this may vary from the stocking stitch if both are used in the same pattern.

Fair Isle technique: When two or three colours are worked repeatedly across a row, strand the yarn not in use loosely behind the stitches being worked. If you are working with more than two colours, treat the "floating" yarns as if they were one yarn and always spread the stitches to their correct width to keep them elastic. It is advisable not to carry the stranded or "floating" yarns over more than three stitches at a time, but to weave them under and over the colour you are working. The "floating" yarns are therefore caught at the back of the work.

SLIP STITCH EDGINGS

When a row end edge forms the actual finished edge of a garment, you will often find a slip stitch edging is worked along this edge. To work a slip stitch edging at the end of a right side row, work across the row until there is one stitch left on the left needle. Pick up the loop lying between the needles and place this loop on the right needle. Please note that this loop does NOT count as a st and is not included in any stitch counts. Now slip the last stitch knitwise with the yarn at the back of the work. At the beginning of the next row purl together the first (slipped) stitch with the picked-up loop. To work a slip stitch edging at the end of a wrong side row, work across the row until there is one stitch left on the left needle. Pick up the loop lying between the needles and place this loop on the right needle. Please note that this loop does NOT count as a stitch and is not included in any stitch counts. Now slip the last stitch purlwise with the yarn at the front of the work. At the beginning of the next row knit together through the back of the loop the first (slipped) stitch with the picked-up loop.

FINISHING INSTRUCTIONS

After working for hours knitting a garment, it seems a great pity that many garments are spoiled because so little care is taken in the pressing and finishing process. Follow the following tips for a truly professional-looking garment.

PRESSING

Block out each piece of knitting and, following the instructions on the ball band, press the garment pieces, omitting the ribs.

Take special care to press the edges, as this will make sewing up both easier and neater. If the ball band indicates that the fabric is not to be pressed, then covering the blocked out fabric with a damp white cotton cloth and leaving it to stand will have the desired effect. Darn in all ends neatly along the selvage edge or a colour join, as appropriate.

STITCHING SEAMS

When stitching the pieces together, remember to match areas of colour and texture very carefully where they meet. Use a special seam stitch, such as backstitch or mattress stitch, for all main knitting seams and join all ribs and neckband with mattress stitch, unless otherwise stated.

CONSTRUCTION

Having completed the pattern instructions, join left shoulder and neckband seams as detailed above. Sew the top of the sleeve to the body of the garment using the method detailed in the pattern, referring to the appropriate guide:

Straight cast-off sleeves: Place centre of cast-off edge of sleeve to shoulder seam. Sew top of sleeve to body, using markers as guidelines where applicable.

Square set-in sleeves: Place centre of cast-off edge of sleeve to shoulder seam. Set sleeve head into armhole, the straight sides at top of sleeve to form a neat right-angle to cast-off stitches at armhole on back and front.

Shallow set-in sleeves: Place centre of cast-off edge of sleeve to shoulder seam. Match decreases at beginning of armhole shaping to decreases at top of sleeve. Sew sleeve head into armhole, easing in shapings.

Set-in sleeves: Place centre of cast-off edge of sleeve to shoulder seam. Set in sleeve, easing sleeve head into armhole. Join side and sleeve seams.

Slip stitch pocket edgings and linings into place. Sew on buttons to correspond with buttonholes.

ABBREVIATIONS

KNITTING ABBREVIATIONS

alt	alternate
beg	begin(ning)
cm	centimetre(s)
cont	continu(e)(ing)
dec	decreas(e)(ing)
foll	following
g st	garter stitch
in	inch(es)
inc	increas(e)(ing)
K	knit
m	metres
M1	make one stitch by picking up horizontal loop before next stitch and knitting into back of it
M1P	make one stitch by picking up horizontal loop before next stitch and purling into back of it
meas	measures
mm	millimetre(s)
P	purl
patt	pattern
psso	pass slipped stitch over
p2sso	pass 2 slipped stitches over
rem	remain(ing)
rep	repeat
rev st st	reverse stocking stitch (1 row K, 1 row P)
RS	right side
skpo	slip 1, knit 1, psso
sl 1	slip one stitch
st(s)	stitch(es)
st st	stocking stitch (1 row K, 1 row P)
tbl	through back of loop(s)
tog	together
WS	wrong side
yd	yards
yfrn	yarn forward round needle
yfwd	yarn forward
yon	yarn over needle
yrn	yarn round needle
0	no stitches, times or rows – no stitches, times or rows for that size

CROCHET ABBREVIATIONS

Simple crochet has been used for edgings for a few garments in this book. British terminology has been used in the pattern instructions. The US equivalents are given below:

UK		US	
ch	chain	ch	chain
dc	double crochet	sc	single crochet
ss	slip stitch	slip st	slip stitch

ABOUT THE YARNS

Rowan Scottish Tweed Yarns (formerly known as Harris Yarns) are available in four weights:
4 ply, DK, Aran and Chunky. The full range of colours is obtainable in 4 ply. The other weights come in some but not all of these colours. All the yarns are to be hand washed or can be dry cleaned at the cleaner's discretion.

Rowan Scottish Tweed 4 ply

A lightweight 100 per cent pure wool yarn
Ball size: 25g ($^7/_8$oz); about 110m (120yd) per ball
Recommended tension: 26-28 sts and 38-40 rows to 10cm (4in) using needle size 3-3$^1/_4$mm/11-10 UK (US sizes 2-3)

Rowan Scottish Tweed DK

A medium weight 100 per cent pure wool yarn
Ball size: 50g (1$^3/_4$oz); about 113m (123yd) per ball.
Recommended tension: 20-22 sts and 28-30 rows to 10cm (4in) using needle size 4mm/8 UK (US size 6)

Rowan Scottish Tweed Aran

A thick 100 per cent pure wool yarn
Ball size: 100g (3$^1/_2$oz); about 170m (186yd) per ball
Recommended tension: 16 sts and 23rows to 10cm (4in) using needle size 5-5$^1/_2$mm/6-5UK (US sizes 8-9)

Rowan Scottish Tweed Chunky

A chunky 100 per cent pure wool yarn
Ball size: 100g (4oz); about 100m (109yd) per ball.
Recommended tension: 12 sts and 16 rows to 10cm (4in) using needle size 8mm/0 UK (US size 11)

OVERSEAS DISTRIBUTORS

AUSTRALIA
Australian Country Spinners,
314 Albert Street, Brunswick, Victoria 3056.
Tel: (03) 9380 3888

BELGIUM
Pavan, Meerlaanstraat 73, B9860 Balegem (Oosterzele)
Tel: (32) 9 221 8594
pavan@pandora.be

CANADA
Diamond Yarn,
9697 St Laurent, Montreal, Quebec, H3L 2N1.
Tel: (514) 388 6188
Diamond Yarn (Toronto),
155 Martin Ross, Unit 3, Toronto, Ontario,M3J 2L9.
Tel: (416) 736 6111
diamond@diamondyarn.com
www.diamondyarns.com

FRANCE
Elle Tricot, 8 Rue du Coq, 67000 Strasbourg.
Tel: (33) 3 88 23 03 13
elletricot@agat.net.
www.elletricote.com

GERMANY
Wolle & Design,
Wolfshovener Strasse 76, 52428 Julich-Stetternich.
Tel: (49) 2461 54735.
Info@wolleunddesign.de.
www.wolleunddesign.de

HOLLAND
de Afstap, Oude Leliestraat 12, 1015 AW Amsterdam.
Tel: (31) 20 6231445

HONG KONG
East Unity Co Ltd, Unit B2, 7/F Block B,
Kailey Industrial Centre, 12 Fung Yip Street, Chai Wan.
Tel: (852) 2869 7110 Fax (852) 2537 6952
eastuni@netvigator.com

ICELAND
Storkurinn, Laugavegi 59, 101 Reykjavik.
Tel: (354) 551 8258
Fax: (354) 562 8252
malin@mmedia.is

JAPAN
Puppy Co Ltd,
T151-0051, 3-16-5 Sendagaya, Shibuyaku, Tokyo.
Tel: (81) 3 3490 2827
info@rowan-jaeger.com

KOREA
De Win Co Ltd,
Chongam Bldg, 101, 34-7 Samsung-dong, Seoul.
Tel: (82) 2 511 1087.
knittking@yahoo.co.kr. Www.dewin.co.kr

My Knit Studio,
(3F) 121 Kwan Hoon Dong, Chongro - ku, Seoul,
Tel: (82) 2 722 0006.
myknit@myknit.com

NORWAY
Paa Pinne, Tennisvn 3D, 0777 Oslo.
Tel: (47) 909 62 818
design@paapinne.no www.paapinne.no

SPAIN
Oyambre, Pau Claris 145, 80009 Barcelona.
Tel: (34) 670 011957.
comercial@oyambreonline.com

SWEDEN
Wincent, Norrtullsgatan 65, 113 45 Stockholm.
Tel: (46) 8 33 70 60
wincent@chello.se www.wincent.nu

U.S.A.
Rowan USA, c/o Westminster Fibers Inc,
4 Townsend West, Suite 8, Nashua, New Hampshire 03063
Tel: (1 603) 886 5041/5043.
rowan@westminsterfibers.com

For all other countries: please contact Rowan for stockists details.

DENMARK
AALBORG: Designvaerkstedet, Boulevarden 9, 9000.
Tel: (45) 9812 0713
Fax: (45) 9813 0213
AARHUS: Ingerís, Volden 19, 8000
Tel: (45) 8619 4044
KOBENHAVN K: Sommerfuglen, Vandkunsten 3, 1467
Tel: (45) 3332 8290
mail@sommerfuglen.dk www.sommerfuglen.dk
KOBENHAVN K: Uldstedet, Fiolstraede 13, 1171.
Tel/Fax: (45) 3391 1771
LYNGBY: Uldstedet, G1. Jernbanevej 7, 2800.
Tel/Fax: (45) 4588 1088
ROSKILDE: Garnhoekeren, Karen Olsdatterstraede 9, 4000.
Tel/Fax: (45) 4637 2063

NEW ZEALAND
AUCKLAND: Alterknitives, PO Box 47961, Ponsonby
Tel: (64) 9 376 0337
knitit@ihug.co.nz
LOWER HUTT: Knit World, PO Box 30 645,
Tel: (64) 4 586 4530
knitting@xtra.co.nz
TAUPO: The Stitchery,
Shop 8, Suncourt Shopping Centre, 1111 Taupo
Tel: (64) 7 378

ACKNOWLEDGMENTS

We would like to thank the following for their help with this book:

John Heseltine for photography (and Tara Heseltine for assisting), Anne Wilson for design, and Emma Freemantle for styling; Penny Hill, Eva Yates and their teams for knitting; Sue Whiting, Stella Smith and Marilyn Wilson for pattern writing and checking, and Sally Harding for proof-reading.

Thanks also to Kate Buller, Ann Hinchcliffe and Lee Wills at Rowan Yarns.

We are also grateful to JJ Locations (and Gloria Thompson) for locations; Models One and Storm for the models, Hannah and Bradley; and Astrid & Alice in Notting Hill, London, for lending garments for photography.